Sustainable Supply Chai

Balkan Cetinkaya • Richard Cuthbertson •
Graham Ewer • Thorsten Klaas-Wissing •
Wojciech Piotrowicz • Christoph Tyssen

Sustainable Supply Chain Management

Practical Ideas for Moving Towards Best Practice

Balkan Cetinkaya
Berlin Institute of Technology (TU-Berlin)
Department of Technology and Management
Chair of Logistics
Strasse des 17. Juni 135
10623 Berlin
Germany
cetinkaya@logistik.tu-berlin.de

Graham Ewer
European Logistics Association
Head Office
Kunstlaan/Avenue des Arts 19
1210 Brussels
Belgium
ela@elalog.org

Wojciech Piotrowicz
University of Oxford
Saïd Business School
Park End Street
OX1 1HP Oxford
United Kingdom
Wojciech.Piotrowicz@sbs.oxford.ac.uk

Richard Cuthbertson
University of Oxford
Saïd Business School
Park End Street
OX1 1HP Oxford
United Kingdom
Richard.Cuthbertson@sbs.ox.ac.uk

Thorsten Klaas-Wissing
Chair of Logistics Management
University of St.Gallen
Dufourstrasse 40a
9000, St. Gallen
Switzerland
thorsten.klaas@unisg.ch

Christoph Tyssen
Chair of Logistics Management
University of St.Gallen
Dufourstrasse 40a
9000, St. Gallen
Switzerland
christoph.tyssen@unisg.ch

ISBN 978-3-642-12022-0 e-ISBN 978-3-642-12023-7
DOI 10.1007/978-3-642-12023-7
Springer Heidelberg Dordrecht London New York

Cover design: WMXDesign GmbH, Heidelberg, Germany

Printed on acid-free paper

Springer is part of Springer Science+Business Media (www.springer.com)

Foreword

The European Logistics Association, ELA, has been engaged with the bestLog project since its inception in 2005. We have always seen its aims as being close to our own core purposes as an association of professional bodies for logistics practitioners across Europe. It is exciting to see the project coming to fruition – doubly so for ELA since we will play a key role in the operation of the Platform which will run on after the Project to promote and enable the sharing of best practice in logistics. The goal of achieving greater efficiency through best practice to achieve better commercial performance with reduced environmental and social impacts is a prize of enormous importance and value to every one of us. This objective is all the more important in the face of the significant changes that have taken place in technology, in society and in the conduct of business in recent years. We now live in a world where almost anyone, and anything, can be connected, and where technology is an integral part of our daily lives. The traditional boundaries in logistics and transport have disappeared to be replaced by an infinitely flexible and seamless process, with logistics and supply chain management as an integrator at the centre of a revolution.

This textbook represents an important part of reaching this goal of greater efficiency through best practice. It draws on the experience and the practical examples of best practice that have been at the centre of the bestLog project research. It is essentially practical in nature. It is aimed squarely at the busy logistics practitioner who wants and needs to know more about his or her profession, and those instructing them. It will provide an important source of reference for everybody, as well as a conduit to other information of a like kind. It is often said that there is no monopoly of wisdom, and there is a lot of wisdom here for us all to use.

It will not surprise you that ELA has had a strong hand in the preparation of this publication. Although directly related to the work of the Project, it is thoroughly in keeping with the spirit of ELA's other work in the field providing definition to the practice of logistics and the Supply Chain. From our perspective, not only will it stand in its own right, but it will join a wider family of reference work that is essential for the exercise of effective logistics. ELA will be playing its part in distributing it, including via the new ELAbestLog Platform http://www.elabestlog.org.

I commend this book to you: it is both timely and relevant.

Dr. Alfonz Antoni
President of the European Logistics Association

Preface

The importance of logistics has increased enormously leveraged by technically effective and relatively low cost means of transport. As a severe consequence this development has led to tremendously increased traffic volumes, which have risen even faster than economic growth in recent decades. The decoupling of this trend has been a major aim of the European Union (EU) policy, and has been controversially discussed in the professional world. In order to get a better understanding of the interrelation between economically driven logistics decisions of single companies and the resulting transportation development, the European Commission (EC) supported a number of research projects concerning these issues. As a result, considerable findings and useful advice have been produced in these projects, which unfortunately were not always made explicit to a wider professional audience. However, to achieve the EC's political objectives with regard to sustainable economic development, these findings must be transferred and translated into the daily operations of (industrial) shippers and logistics service providers. Therefore, a project called bestLog (http://www.bestlog.org) was initiated in 2005/2006 consisting of nine partners from different European countries (Belgium, Czech Republic, France, Germany, Poland, Spain, Sweden, Switzerland, United Kingdom). Amongst others, one major goal of bestLog is the promotion and dissemination of logistics best practice into the professional community of logisticians and supply chain managers. This has been done by multiple means, including dedicated meetings, workshops, and conferences. Through such events the topic of sustainability in logistics and transportation, i.e. solutions which are economically, socially and environmentally favourable, became more and more evident.

Constituting one major part of the bestLog activities, this textbook about Sustainable Supply Chain Management (SCM) was written, to present and summarise the specific insights and experiences of the bestLog project and to reproduce this knowledge in a structured, comprehensive and accessible manner. Though all contributors have a strong academic background, this book intentionally does not aim to be overly scientific. Instead, it is written for professionals and trainers striving for logistics and supply chain management know-how from practice.

In order to integrate as much conceptual knowledge and practical experience as possible into this book we divided this book into four sections:

The first part of this volume (Part I) deals with the fundamental need for sustainable SCM in the light of changing customer demands and the current developments in logistics and transportation.

The second part (Part II) takes on the need for sustainable SCM and provides different levers and areas of action for sustainable SCM. These areas are derived from the specific experiences of the bestLog project and structured in order to tackle certain areas of sustainable SCM. In order to make the insights as explicit as possible, we integrated little illustrative hints in boxes within the text. Those hints are derived from the real world bestLog-cases.

Part III provides an outlook on future developments in Supply Chains.

Part IV then closes the book with the case studies used for illustration in Part II. The listing of the cases follows a certain structure in order to provide a quick overview and allow the interested reader to quickly elect the right case for their needs.

Of course, the completion and publishing of this book would not have been possible without the assistance and contribution of many people. First of all we thank the entire bestLog-team for four years of intense work and "sustainable" progress. Also we would like to thank representatives of companies and institutions which supported our work providing data for case studies and shared their experience during project workshops and conferences. Furthermore we thank the reviewers who supported us with expertise and by submitting valuable recommendations to the authors. Here we especially thank the members of the Advisory and Communications Board of bestLog with their fruitful contributions and input (Please see the following table for the names and institutions). Additionally, we thank Springer for being such a supportive publisher. For their administrative and secretarial support, we additionally thank Maria Horn, Manuel Rusch, Katrin Schuler, Marcel Tulke and Florian Wimmer. Last but not least, we would especially like to thank the European Union for their financial support of such an important project!

Berlin, Germany — Balkan Cetinkaya
St. Gallen, Switzerland — Thorsten Klaas-Wissing
Oxford, United Kingdom — Richard Cuthbertson
Oxford, United Kingdom — Wojciech Piotrowicz
Brussels, Belgium — Graham A. Ewer
St. Gallen, Switzerland — Christoph Tyssen

Endorsement

The volume of freight transport in Europe is growing constantly, and this has obvious disadvantages in terms of congestion, safety risks and pollution. This creates major problems for the supply chain performance of many companies and is perceived as a threat to the competitiveness of European production and retail systems and therefore also future growth.

These developments have been in the focus of the European transport policies geared towards creating efficient supply chains and co-modality – the efficient use of different modes of transport on their own and in combination – with the aim of ensuring optimum and sustainable use of resources.

Four years ago, the European Commission's Directorate General for Transport and the Berlin Institute of Technology launched one of the most wide-ranging research, promotion and information projects in the field of European logistics and transport – the bestLog project – in response to this far-reaching challenge.

BestLog was launched to facilitate better alignment between society, environment and business – and to foster the efficient and sustainable use of resources in the area of transport and logistics. Together with nine universities, logistics consulting firms from across Europe and the ELA, bestLog has defined, collected and disseminated case studies of good logistics practices, established a platform for ongoing exchange on good logistics and supply chain practices, developed a certification scheme for sustainable logistics and supply chain management, published regular reports on the state of the art in European logistics education, developed training and education packages and published this book on sustainable logistics and supply chain management.

This book focuses on the need to develop supply chains that are sustainable in economic, environmental and societal terms. It incorporates the results of our research and the countless discussions with hands-on practitioners, industry experts, trade association representatives, academics and politicians about good sustainable SCM, logistics and transport practices, their strategies, their formulas, their challenges, their drivers and their measurement.

As a result, this book does not claim to be highly scientific. It is rather designed to render transparent the critical elements and issues of sustainable supply chains that are the key to the future of competitive supply chains in retail and production systems in big companies and SMEs.

I hope that the practical insights and success strategies we have identified will help you to "find your bearings" and to adapt your own supply chain and logistics strategy as and when needed.

I would like to thank all the individuals and companies involved for their extremely active participation in this project. I am particularly grateful to the 32 members of our European Advisory and Communication Board who were involved from the beginning and to the nine bestLog team partners: the European Logistics Association, the University of Oxford – Saïd Business School, the University of St. Gallen – Chair of Logistics Management, the Chalmers University of Technology, the Warsaw School of Economics – Logistics Department, ITENE – Packaging, Transport and Logistics Institute, Reliant s.r.l. and Mettle.

I would like to thank also the European Logistics Association (ELA) who decided to continue with the platform after the end of the project via the new ELAbestLog Platform. ELA's unique network of national member associations covering more than 30 European countries will be a success driver to promote the exchange platform across Europe.

Last but not least, I would like to thank the European Commission for the excellent cooperation and for their trust in our expertise as a longstanding scientific partner.

Looking forward towards the next edition we would welcome your criticisms, praise your suggestions. I hope this book makes for a stimulating and interesting read.

Frank Straube
Managing Director
Chair of Logistics
Department of Technology and Management
Berlin Institute of Technology (TU-Berlin)

Contents

Part I
Section A

Chapter 1
The Need for Sustainable Supply Chain Management

Richard Cuthbertson

1.1 Sustainable Supply Chains

This book focuses on the need to develop sustainable supply chains – economically, environmentally and socially. This book is not about a wish list of impractical choices, but the reality of decisions faced by all those involved in supply chain management today. Our definition of sustainable supply chains is not restricted to so-called "green" supply chains, but recognises that in order to be truly sustainable, supply chains must operate within a realistic financial structure, as well as contribute value to our society. Supply chains are not sustainable unless they are realistically funded and valued. Thus, a real definition of sustainable supply chain management must take account of all relevant economic, social and environmental issues.

1.2 Best Practice in Supply Chain Management

Supply chain management is the management of a network of interconnected businesses involved in the ultimate provision of product and service packages required by end customers (Harland 1996). Thus, supply chain management covers all the necessary movement and storage of raw materials, work-in-process inventory, and finished goods from the point-of-origin to the point-of-consumption.

While transporting goods to where they are most valued makes economic sense, there are major social and environmental implications. For example, trucking product by road may lead to congestion and pollution. However, by packing a vehicle more densely, we can reduce both congestion and pollution, thus creating a more sustainable supply chain. This book is the culmination of a 3-year worldwide search for "best" practices in logistics and supply chain management to do just that – to develop more sustainable supply chains.

C. Tyssen et al., *Sustainable Supply Chain Management*,
DOI 10.1007/978-3-642-12023-7_1,

It is worth noting that "best" practices are both relative and transitory. They are neither "best" in every way, nor "best" for all time. They are the best practice within the specific context at that moment in time. Hence, the case studies do not attempt to provide universal best practices, but instead provide specific examples of best practice, which universally might be considered good, successful or improved approaches to the case in point.

While best practices will not necessarily transfer from one situation to another, there is learning to be found everywhere. Today, supply chains know no boundaries and learning from other people's successful practices from other companies, supply chains, industries and countries is hugely valuable. This book contains examples from a wide range of situations. Moreover, to accumulate and synthesize learnings from these many different situations allows frameworks to be developed that provide the fundamental building blocks at the centre of successful logistics and supply chain management. These learnings are clearly identified throughout the text and summarised in the form of learning goals, questions and further reading at the end of each chapter.

1.3 The Need for Sustainable Supply Chains

The work behind this book has been funded by the European Commission. The European Commission recognise the importance to the economy of efficient and effective supply chains, and also recognise the potential negative impacts on society and the environment – hence, their funding of this work.

Freight transport can be considered to be the key element in modern supply chains. The demand for freight transport continues to increase. This increase is due not only to the increasing demand for physical products, but also to many supply chain management decisions. As supply chains have become more sophisticated, they have sought to become more responsive, more reliable and more efficient, often leading to increased transport requirements. This was highlighted in the SULOGTRA report to the European Commission in 2002, as discussed below.

The application of modern management principles throughout the supply chain, such as Just-In-Time, lean production and Efficient Consumer Response, have led to more responsive, more flexible supply chains that enable firms to compete in a global market.

The wider geographical sourcing of supplies and wider distribution of finished products is extending supply chains, both upstream and downstream. Global sourcing strategies are driven by the potential savings in purchasing, manufacturing and/or labour costs. This trend is supported by the ease with which information can now flow between different firms throughout the world in real-time. This trend towards global rather than local supply chains is also a result of companies focusing on their core competencies. The vertical disintegration of production allows companies to increase quality of product and service while reducing risk.

The spatial concentration of production has resulted in a reduction of the total number of factories. The focus in many sectors has been to move away from nationally-based production towards single locations producing a particular product for the whole of a continent, or even the world market. This has increased the demand for freight transport. By rationalising their supply base, companies also lower their overall transaction cost at the expense of increasing the logistics and transport services needed to deliver the material flow. This concentration of suppliers is particularly utilised in large scale industries, where the producer has a large degree of power within the supply chain, for example in the automotive industry. Similarly, the spatial concentration of inventory has resulted in a reduced number of stockholding points. These can yield a large financial benefit, outweighing the cost to the firm of any additional transport required, but what are the implications for the environment and society?

Reverse logistics has been helping to alleviate some of the environmental and societal concerns, for example through more efficient backhauling and greater recycling capability. However, reverse logistics may increase transport demand and potentially reduce utilisation rates. Similarly, while technology has reduced the need for regular mail through the development of electronic mail services, the demand for small units, such as parcels, has increased due to online retailing activity and customer-to-customer transaction websites, such as EBay – thus increasing freight transport requirements.

All of the above, especially the underlying need for flexibility, have impacted the freight modal split, with the result that road transport has continued to take an increasing share of the expanding demand for freight transport. Changing cost structures, better infrastructure, the standardisation of loading units and deregulation have all played their part in supporting this trend. The SULOGTRA research project summed up these trends, as the following table demonstrates (Table 1.1).

Given these developments, the driving force behind this book is the continued growth of transport requirements, especially road transport, in a world of finite resources. Greater transport operations foster economic growth and are an essential element in maintaining and increasing the competitive advantage of companies and nations, yet transport requirements also create social and environmental costs and benefits. This book aims to encourage the positive elements of supply chain management while decreasing the negative impacts.

1.4 The Implications of Modern Supply Chain Management

While actual figures are always subject to change, update, and arguments over definitions, the trends in modern supply chain management are clear. The following chart shows freight transport growth of the EU27 countries in terms of weight (tonnes) and weight carried by distance (tonne km), as well as economic development (measured in GDP at constant prices) (Fig. 1.1).

While the growth in freight transport may be a concern due to the potential negative social and environmental impacts, the growth relative to GDP is even more of a concern because it is a more recent and more discriminate development.

Table 1.1 Key transport trends (Source: SULOGTRA)

Trends	Transport demand
Restructuring of logistics network	
Spatial concentration of productionProduction has become increasingly concentrated, e.g. through mergers and acquisitions, as companies try to streamline production and to gain synergies.	Up
Spatial concentration of inventoryWarehousing and stockholding costs have lead companies to centralise their warehousing in few or one locations. This leads to longer distances from sources to destinations.	Up
Development of break-bulk/transhipment systemsTranshipment points, while often centralised (see above), offer the potential for better utilisation of trucks. This trend is significant especially in retail.	Down
Reverse logisticsPartly due to legal requirements but also because of potential savings goods and packaging are increasingly returned to the manufacturer for recycling. This leads to higher transport demand and in many cases lower average load factors.	Up
Realignment of supply chains	
Rationalisation of the supply baseSupplier reduction leads to longer transport distances in procurement.	Up
Vertical disintegration of productionWhen companies specialise on core competencies, this implies more handling steps from raw material to finished product and hence larger transport demand.	Up
Wider geographical sourcing of suppliesLike supplier reduction, global sourcing tends to lead to longer transport distances.	Up
Postponement – Deferred customisationThe further the customisation of products is deferred, the later variants of a product are produced the better will be load factors. On the other hand there may be a tendency towards longer distances travelled as the logistics structure is becoming more decentralised.	Up
Increased direct deliveryDirect delivery tends to decrease distances travelled while at the same time reducing load factors.	Up
Rescheduling of product flows	
Time-compression principlesJIT, Lean, ECR and the like, all target the reduction of inventory but at the same time tend to reduce load factors.	Up
Growth of "nominated day" deliveries and timed delivery 0systemsThese lead to reduced load factors.	Up
Changes in product design	
Product modularisationModules are usually bulkier than single items, thereby causing more transport demand.	Up

This growth in freight transport can be measured relative to GDP to provide an elasticity measure of freight transport.

$$\text{Freight transport elasticity} = \frac{\text{Growth in freight transport measured in tonnekm}}{\text{Growth in GDP measured in relevant currency}}$$

While freight transport measured in tonne km has grown by +185% since 1970, there has been an important change in the nature of that growth. Before 1980, transport volumes grew in line with the economy, thus having an elasticity of tonne

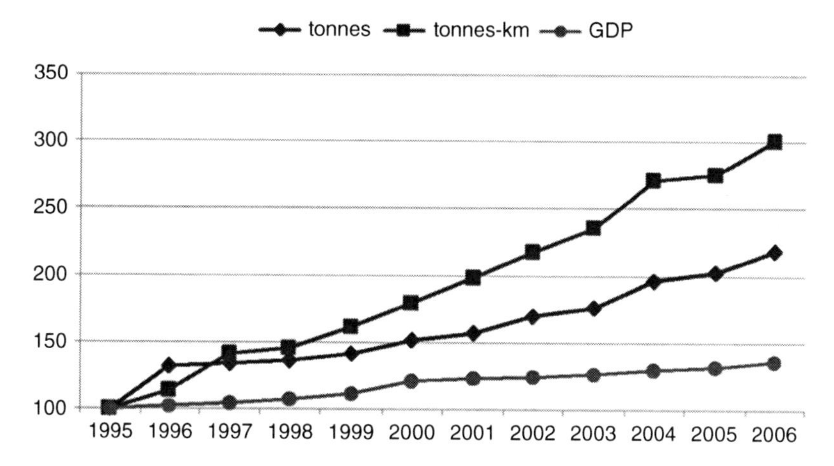

Fig. 1.1 EU27 Freight transport growth 1995–2006

	Growth GDP	Growth Total Freight 1995-2004	Elasticity			Growth Freight w/o International Road 1995-2004	
Spain	4,23	11,45	2,71		2,17	9,19	Spain
Estland	8,01	21,00	2,62		1,85	17,42	Ireland
Slovenia	4,53	10,62	2,34		1,60	7,41	Greece
Ireland	10,56	20,83	1,97		1,51	5,13	Slovenia
Portugal	2,84	5,25	1,85	> 1	1,27	2,76	Denmark
Austria	2,47	4,10	1,66		1,14	8,92	Estland
Greece	4,64	7,49	1,62		1,11	3,15	Portugal
Lithuania	7,21	10,62	1,47		1,06	2,81	The Netherlands
Luxembourg	5,80	7,28	1,26		0,89	2,20	Austria
The Netherlands	2,65	3,32	1,25		0,89	7,60	Lithuania
Germany	2,16	2,42	1,12		0,84	2,14	France
Latvia	8,48	8,15	0,96		0,72	1,56	Germany
Italy	1,50	1,25	0,84		0,62	1,46	Belgium
Poland	5,09	3,55	0,70		0,61	1,83	Poland
Hungary	4,86	3,36	0,69		0,50	2,14	Finland
Finland	4,30	2,71	0,63	0 < x < 1	0,44	1,35	Sweden
Denmark	2,16	1,24	0,58		0,42	3,57	Latvia
France	2,54	1,35	0,53		0,35	0,52	Italy
Sweden	3,06	1,49	0,49		0,33	1,06	United Kingdom
Czech Republic	2,33	1,13	0,49		0,16	0,91	Luxembourg
Belgium	2,35	1,08	0,46		0,02	0,10	Hungary
United Kingdom	3,25	0,91	0,28		0,00	0,00	Malta
Malta	0,93	0,00	0,00		-0,15	-0,48	Czech Republic
Slovac Republic	4,82	-0,79	-0,16	< 0	-0,53	-2,59	Slovac Republic
Cyprus	3,95	-0,75	-0,19		-1,10	-3,48	Cyprus
			1,05	Average	0,66		

Fig. 1.2 Freight transport growth by country and GDP

km to GDP close to unity, or 1. It is only since the 1980s, that we have seen freight transport growing at a faster rate than the overall European economy.

Furthermore, this overall trend hides some important differences across the different regions of Europe. The following table shows the transport elasticity by country in order to evaluate the relationship between freight growth and GDP growth (Fig. 1.2):

The table shows that the elasticity of total freight based on GDP (at constant prices) in Europe from 1995 until 2004 is on average 1.05. This means that a 1.00% increase of GDP leads to a 1.05% increase of total freight, so freight transport is growing 5% more than GDP. However, this progressive relationship is only valid for 11 out of 25 countries. Note that the clustering of these countries is not explained by any simple geographical clustering, such as Eastern or Western Europe.

In order to understand the effect of International Trade on this relationship, international road transport (which is the dominant transport mode in almost every country in Europe; measured by cross-trade and cabotage) was subtracted from the total transport. The elasticities of each country and the average elasticity of 0.66 point out that there is a perceptible increase in the elasticity of freight transport with respect to economic activity from 1994 until 2005, which can be explained to a large extent by the rapid increase in international transport. The statistical clustering of these countries again is not explained by any regional clustering.

Hence, while there is a trend of coupling transport growth with growth in economic activity or trade, there are also clear differences between countries. The differences are based on a combination of several phenomena, which have an impact on the development of the "Transport Demand", of all modes taken together.

It is worth noting that measuring freight demand is difficult. In particular, the standard measure of tonne km does not clearly identify changing volume/weight ratios. If weight is reduced and the distance extended, then the tonne km calculation may stay the same. Moreover, more qualitative aspects are not captured, such as the speed and reliability of delivery. Using GDP as a measurement of economic activity may also be criticised as it includes intangible services as well as physical industrial output. However, there is no denying that recent increases in economic growth appears to be linked with increased transport requirements. Any improvements in supply chain management are therefore desirable to the firm, as well as to the wider community.

More recent figures from the 2009 Road Freight Transport Vademecum show that total freight transport by inland modes in the EU totalled 2,595 billion tonne-kilometres (tkms) in 2006. This represented about 73% of the inland freight transport market, while rail had a share of 17%, with inland waterways and oil pipelines accounting for 5% each. Moreover, sustainable supply chain management is not just an issue for large corporations. The road haulage sector is fragmented and includes a large number of small businesses units, especially in Spain (more than 130,000 companies in 2005), Italy (100,000 companies) and Poland (70,000 companies).

To emphasize that more sustainable supply chain management is not just an issue of demand, it should be noted that the share of empty journeys has increased in recent years, both in national and international transport.

1.5 Moving Towards Sustainable Supply Chains

Transport within a supply chain has many potentially negative social, economic, and environmental impacts. These impacts can be expressed in monetary terms, estimated by calculating direct, indirect and avoidance costs. Studies have placed the external costs of transport, such as pollution, noise, and accidents at 3–5% of

GDP.[1] The results of this study show that road and air have the highest external costs, and rail by far the lowest per transport unit. These external costs are not fully reflected in the price of transport, but are borne by society as a whole.

Externalities of transport include many different social, economic and environmental effects.

For example, from a social perspective, accidents are a major negative impact. Although the number of accidents overall has decreased in developed markets, there are still huge differences across the world. Many countries still have relatively high accident and fatality rates (Fig. 1.3).

From an economic perspective, issues such as congestion have a huge impact on speed of delivery and reliability. Note that congestion is not just restricted to urban areas, but may affect major transport routes, especially where there is a natural obstacle, such as the Alps or the Pyrenees where programmes are underway to improve the infrastructure.

Perhaps, the most discussed impact of transport is in terms of engine exhaust from internal combustion engines, which contains many harmful gases and particles, including carbon dioxide (CO_2). CO_2-Emission per vehicle has decreased on average due to advanced technologies in CO_2-Emission reduction. However, as motor vehicle use increases, the transport shares of such emissions are raised. CO_2 emissions from transport in general and road transport in particular have been rising faster than emissions from all other major sectors of the economy. In 2006, the emissions from

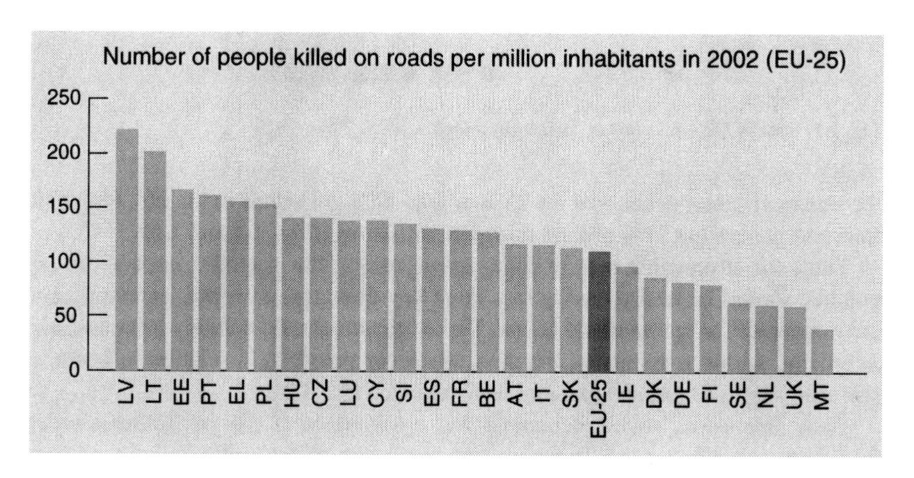

Fig. 1.3 EU25 Road deaths by country
Source: Energy and Transport Report 2000–2004, European Commission)

[1]Kageson (1993), "Getting the Prices Right, A European Scheme for Making Transport Pay its True Costs".

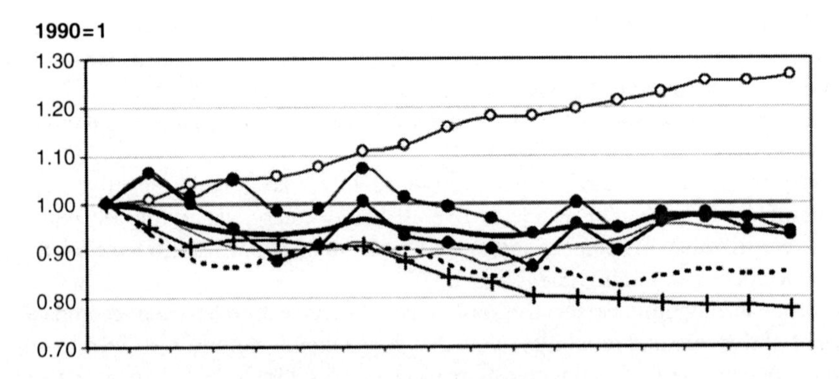

Fig. 1.4 CO2 emissions by sector in EU-27 – 1990–2006

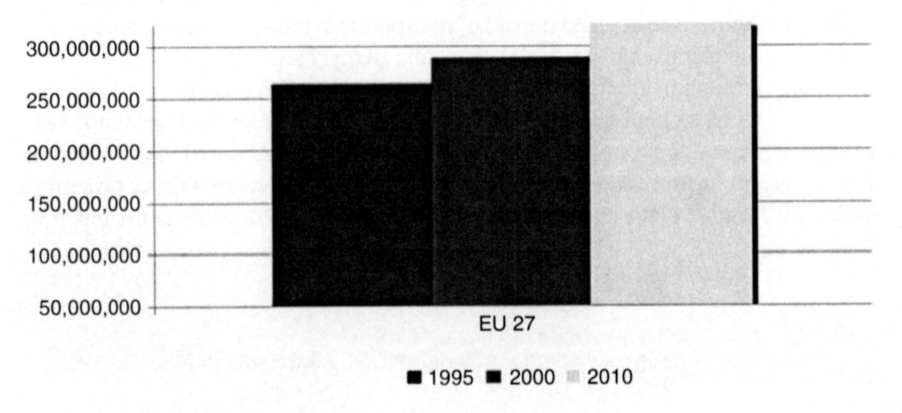

Fig. 1.5 Total CO2 emissions of lorries (in tones) – 1995, 2000, 2010

the transport sector accounted for 23% of total CO_2 emissions in the EU, with road transport generating 71% of total transport emissions (Figs. 1.4 and 1.5).

Thus, for sustainable supply chain management, the potential impacts of any practice should be understood across three key dimensions: social, economic and environmental, as summarized below. These impacts should not only be considered internally, within a company, but also, wherever possible, in relation to business partners, supply chain and external environment (Fig. 1.6).

Each dimension, social, economic and environmental can be further divided into three sub-dimensions. Improvement along any of these dimensions, without a detrimental effect in another dimension, will lead to more sustainable supply chains.

Economic growth has usually led to greater freight transport activity within supply chains. While economic growth obviously continues to be desirable, the associated growth in the negative effects of transport requirements is unsustainable in the long term. Thus, this book aims to reduce the need for continually increasing the negative impacts of freight transport without damaging economic growth. This

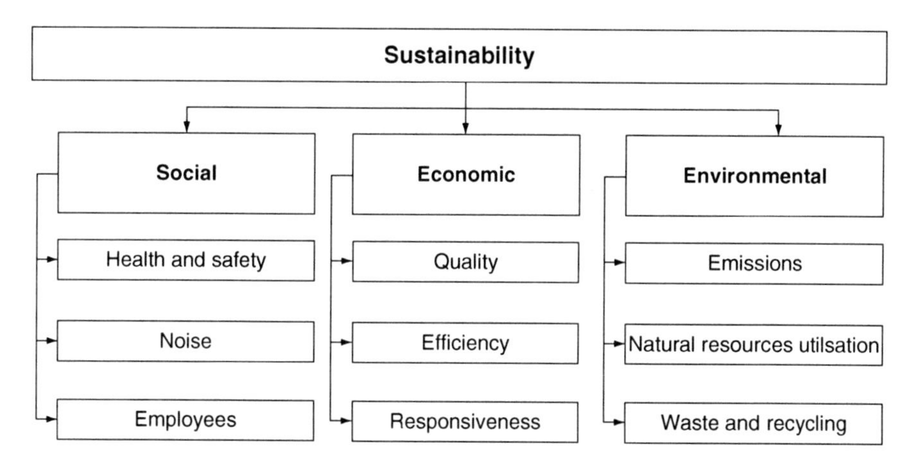

Fig. 1.6 Metric dimensions and sub dimensions

may seem an unattainable goal but, as will be seen in the pages that follow, it is possible to move towards. There are so many different ways that we can improve supply chain management today and for the future. Each of these approaches differs in levels of complexity, technology, and investment, but all require one key ingredient that on its own can make substantial change, namely people and their attitudes towards: managing effects rather than ignoring impacts, increasing education rather than assuming insularity, and improving standards rather than minimising expectations.

1.6 The Structure of This Book

This book can be read in full, however the Chapters may also be read independently, so it is possible for readers to concentrate on their main topic of interest. An alternative approach is to read the practical case studies: supply chain best practices, and then move to the main text.

Each chapter includes tasks or questions. They provide readers with a chance to apply their learning from the knowledge presented in the text, while drawing on their own experience. It is important to stress that answers to the tasks are strongly dependent on the context assumed. Different answers in different contexts may highlight particular supply chain complexities.

Part II focuses in on more detail of the general findings from the Bestlog project. The starting point is the strategic role of supply chain management within the business (Chap. 2), which is usually focused on understanding the economic goals of cost efficiency, customer responsiveness and overall quality. In order to ensure that the chosen supply chain strategy is both achievable and sustainable, there is then a need to consider the overall impact of the operations, by measuring and monitoring the economic, social and environmental consequences, both positive and negative (Chap. 3). To be successfully implemented, this requires an integrated approach to the management of products, processes, knowledge and infrastructure (Chap. 4). The Bestlog findings continually emphasised the importance of successfully managing people and teams in order to continually work towards creating sustainable supply chains. There is also a need to influence the requirements and expectations of stakeholders external to the immediate supply chain, and outside the direct control of the organisation (Chap. 5). Finally, during the collection of Bestlog case studies over 3 years, the importance of risk management was highlighted (Chap. 6). There are a number of case studies not included in this book because although they initially appeared to contribute towards sustainable supply chain management, they proved not to be sustainable and were overtaken by events, such as new technology or a new economic environment. While all approaches to supply chain management may eventually evolve, by taking account of the learnings provided in Part II, developments in managing supply chains should become more robust and hence more sustainable.

In summary, the chapters in Part II consider:

- What is a successful supply chain strategy? (Chap. 2)
- How should supply chain performance be measured? (Chap. 3)
- How should the organisation be managed? (Chap. 4)
- How should stakeholders be influenced? (Chap. 5)
- How can unexpected events be incorporated? (Chap. 6)

Within each chapter in Part II, the relevance and importance of the topic to supply chain management is discussed, along with the resulting challenges and benefits. A range of solutions to the issues identified are considered, and provide an understanding of when to implement which solution. The use of case studies as reference points and reality-based examples is used extensively throughout the book. Furthermore, each chapter provides some recommendations for further reading where the reader wishes to follow up particular issues in more detail.

Part II leads the reader through the core analysis and findings of the 28 collected case studies that appear in Part III. These findings have been tested in four industry and policy workshops (London, Amsterdam, Berlin and Valencia) to include further input from practitioners and policy makers on their thoughts and experience regarding the issues concerned. The cases are based on fieldwork (interviews), and fulfil the certain validation requirements for a case to be considered both real (i.e. implemented and operational) and successful (i.e. more sustainable).

Part III provides a selection of key cases, highlighting different issues, the challenges faced, the solutions adopted, the lessons learned, benefits gained, and the potential transferability to other companies and situations. Further cases can be found at the Bestlog website, run by the European Logistics Association.

Finally, Part IV looks ahead to future developments in economic activity and transport needs, and considers what will be the attributes of logistics and supply chain winners and losers in the future.

This approach of this book has been supported through a whole range of other dissemination activities, from a dedicated website (bestlog), through newspaper and journal articles, to workshops, conferences and executive education

Tasks:

1) Why should any individual commercial organisation attempt to manage sustainability rather than just satisfy relevant regulation?
2) How might casualties and CO_2 emissions (see Figures 1.3 and 1.4) influence the public perception of transport providers?
3) Who should cover external costs of transportation? List some possible solutions.
4) Should Supply Chain members, logistics companies, and transport providers take actions to reduce the negative impacts on sustainability, or merely comply with defined standards?

Part II
Section B

Chapter 2
Developing a Sustainable Supply Chain Strategy

Balkan Cetinkaya

Learning Goals.
By reading this chapter you will:

- Know the basics of competitive strategy and supply chain strategy and understand their interrelations
- Understand the need for a sustainable supply chain strategy
- Understand the ingredients of a sustainable supply chain strategy
- Apply a generic, iterative approach to develop your sustainable supply chain strategy
- Apply a balanced scorecard to implement your sustainable supply chain strategy

2.1 Introduction: The Starting Point

Long-term trends pose challenges for supply chain managers and make increasing requirements on the strategic management expertise of today's companies. These trends include ongoing globalisation and the increasing intensity of competition, the growing demands of security, environmental protection and resource scarcity and, last but not least, the need for reliable, flexible and cost-efficient business systems capable of supporting customer differentiation. More than ever, modern supply chain managers are confronted with dynamic and complex supply chains and therefore with trends and developments that are hard to predict.

In years to come, supply chain management will therefore take on additional strategic tasks that extend beyond its current more operational scope of activity. In order to respond to these changes and remain competitive, supply chain managers need to be able to identify and understand new sustainability issues in their company and business environment.

This calls, especially in respect of global, international, and fragmented supply chains, not only for highly efficient supply chain operations, but also for networking skills that must continuously adapt to sustainability demands to create sustainable,

C. Tyssen et al., *Sustainable Supply Chain Management*,
DOI 10.1007/978-3-642-12023-7_2, © Springer-Verlag Berlin Heidelberg 2011

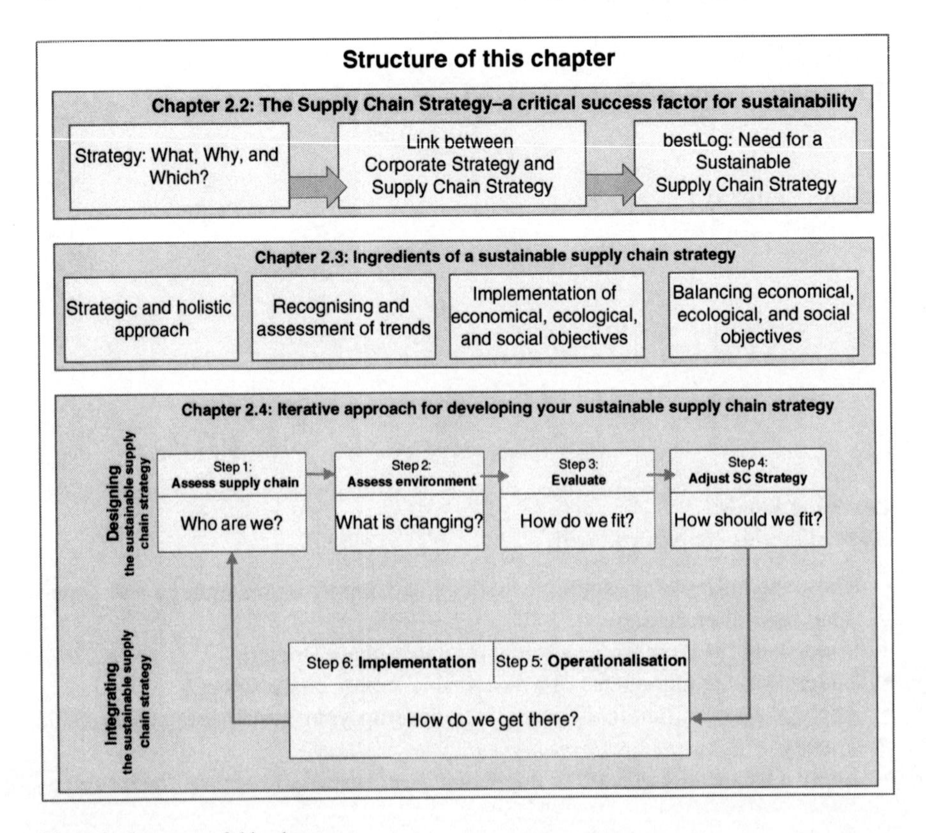

Fig. 2.1 Structure of this chapter

long-term customer-focused supply chains. It calls for the development of sustainable supply chain strategies which create a sustainable competitive advantage.

We will focus in this chapter on how to develop a supply chain strategy for sustainability, and on how to integrate it into your existing supply chain. Section 2.2 introduces the basics of competitive strategy and supply chain strategy, the relationship between the two, and finally, their links with sustainability. Section 2.3 describes the main ingredients of a sustainable supply chain strategy identified in the course of the bestLog project research. Section 2.4 describes a generic, iterative six-step approach to developing and implementing a sustainable supply chain strategy. The following figure illustrates the structure of this chapter (Fig. 2.1).

2.2 The Supply Chain Strategy: A Critical Success Factor for Sustainability

We mentioned earlier the necessity for strategic management in today's supply chains. We will now briefly address the following questions: What does a competitive corporate strategy involve? What is a supply chain strategy, and how is it linked

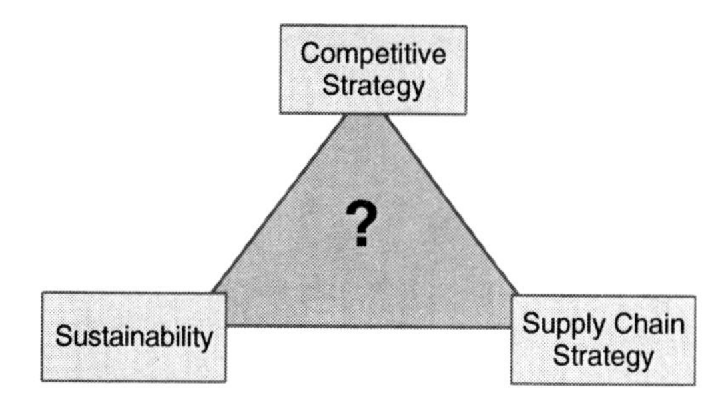

Fig. 2.2 Three areas to integrate

to the competitive strategy? And where is the link to sustainability? Figure 2.2 shows the three areas and their links, which we will try to understand in this section.

The word "strategy" derives from the ancient Greek "strategós" which in turn derives from two words: "stratos" (army) and "ago" (leading). Today's definition of competitive strategy is a holistic, long-term plan for a company[1] to find a distinctive way of competing in order to guarantee profitability over a limited time span, considering the development of its business environment. A competitive strategy is specified by a bundle of aims and objectives to establish a *competitive advantage*,[2] which allows the company to outperform others in the same industry or market. One indicator for this outperforming capability is the company's profitability, compared to the industry average.

According to Porter, there are two basic types of *competitive advantage* a company may pursue: low cost, or differentiation. Porter combines these with the scope of activities that a company seeks to pursue and derives three generic strategies for achieving excellence and market success: cost leadership, differentiation, and focus. The focus strategy has two variants, cost focus and differentiation focus.[3]

In cost leadership, a company aims to become the lowest cost producer in its industry. The sources of cost advantage vary between industries. Usually they include economies of scale, proprietary technology, and preferential access to raw materials.[4] With a differentiation strategy, a company seeks to develop products and services that are perceived as unique in its industry, and which create a value advantage for its customers. This emphasizes the importance of focusing on one or more attributes that customers perceive as important, which usually leads to higher cost levels. But customers of these strongly differentiated companies are loyal to its

[1]Strategy in diversified companies is many-faceted and may be defined at business unit or corporate or company-wide levels. Competitive strategy seeks to create competitive advantage in each of the fields in which a company competes. Corporate strategy concerns two different questions: what business the corporation should be in, and how the corporation should manage an array of business units. See Porter (1987).

[2]Porter (1985).

[3]Porter (1985).

[4]Porter (2008), p. 11.

services and products, are less price-sensitive, and reward the effort made by paying premium prices.

In general we can distinguish – following markets and competition theory – three decisive factors which determine the business environment and consequently the strategy of a corporation: Demand (e.g. customers, target groups, etc.); Supply (e.g. competitors, employees, suppliers, etc.); and the General Environment (e.g. regulations, society, natural resources, etc.). In today's business environment all these three factors are becoming increasingly complex, are changing over time, and together determine the behaviour of market players. Hence a great challenge for strategists is to satisfy on the one hand shareholders, and on the other hand more and more other stakeholders; especially with regard to the latest developments involving sustainability, where certain stakeholders like interest groups for CSR or environmental protection (e.g. Greenpeace) are becoming more and more powerful.

A company's supply chain now plays an important part in the aforementioned three decisive factors and therefore represents an essential strategic resource in the achievement of the strategic goals. For example, customers increasingly recognise the value of supply chain service and quality and are less likely to select products and services only on price. Companies like Apple, Dell, and Procter & Gamble, for example, increasingly outperform others in supply chain excellence.[5] It is now recognised that not only companies but rather, whole supply chains, are in competition.

Both in theory and practice, we find two basic supply chain types, having the potential to assist competitive strategy in the achievement of both cost leadership and differentiation strategy: Lean, cost, efficiency-driven supply chains, and in contrast agile, fast, service driven supply chains.[6]

A lean supply chain fits well with a cost leadership strategy, and is particularly successful, if total logistics costs represent a high proportion of the cost of goods sold, and if the supply chain offers sufficient possibilities for reducing and controlling these costs. An agile supply chain strategy fits well with a differentiation strategy, particularly if customer-oriented differentiation is essential and supply chain solutions need to be segmented and diversified.

Best practice companies do not focus on just one, fixed supply chain strategy. There is an increasing need – and competitive advantage follows – to customise supply chains individually (with regard to different customers, countries, and products) and in consequence to implement multiple supply chain strategies and solutions; especially where quite heterogeneous customer-product mixes need to be supported within the same global supply chains.

The strategic challenge for a supply chain manager is to configure and develop holistically all the multi-layered fields of a supply chain aiming as a whole a strong alignment with the competitive and corporate strategy. The "bridge" from corporate

[5]O'Marah and Hofman (2010).

[6]Christopher (2005), Chopra and Meindl (2004) and Fisher (1997).

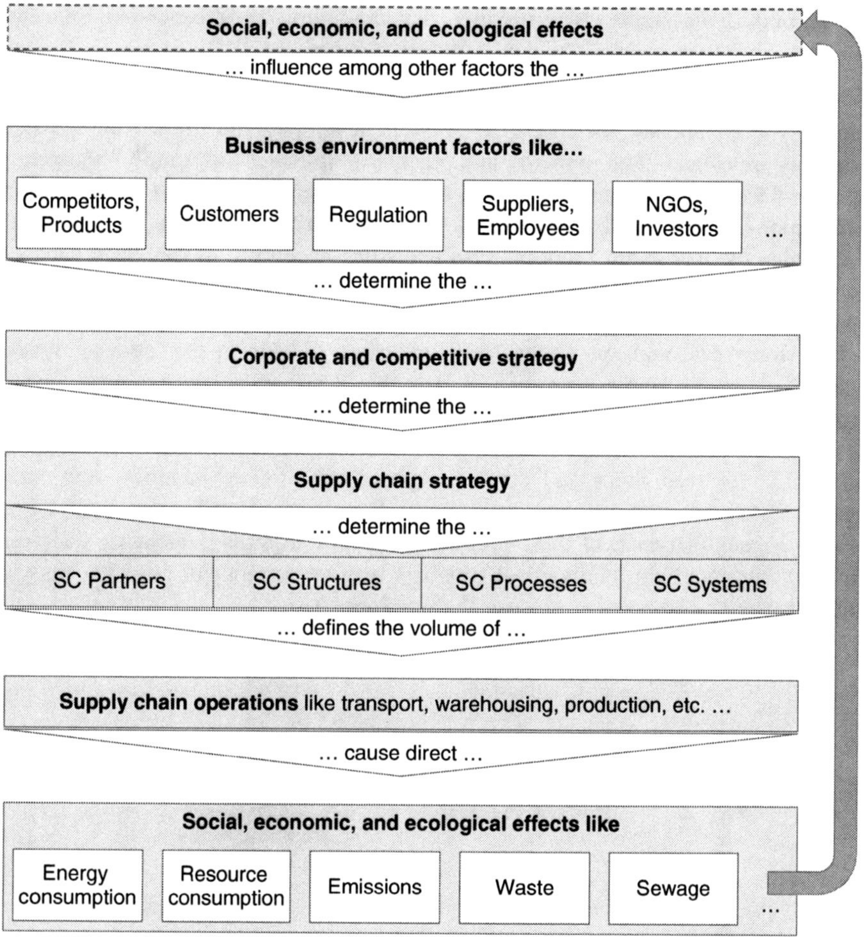

Social, economic, and ecological effects
... influence among other factors the ...
Business environment factors like...
Competitors, Products
Customers
Regulation
Suppliers, Employees
NGOs, Investors
...
... determine the ...
Corporate and competitive strategy
... determine the ...
Supply chain strategy
... determine the ...
SC Partners
SC Structures
SC Processes
SC Systems
... defines the volume of ...
Supply chain operations like transport, warehousing, production, etc. ...
... cause direct ...
Social, economic, and ecological effects like
Energy consumption
Resource consumption
Emissions
Waste
Sewage
...

Fig. 2.3 Supply chain strategy as a bridge between competitive strategy and sustainability

and competitive strategy to supply chain types is the supply chain Strategy. The supply chain Strategy determines the goals and the configuration of the supply chain with regard to supply chain partners, structures, processes, and systems (see Fig. 2.3). In detail these are:

- Regarding the supply chain partners e.g. selection of partners, configuration of outsourcing models and associated gain and cost sharing models
- Regarding the supply chain structures – e.g. configuration of distribution or production network structures in terms of vertical and horizontal stratification
- Regarding the supply chain processes – e.g. configuration of procurement, production, and distribution processes with regard to costs, reliability, speed, and flexibility

- Regarding the supply chain systems – e.g. configuration of leadership, information, reporting, controlling, and incentive systems

In accordance to the aforementioned three decisive factors which determine the strategy of a company, the criteria involved in developing the "right" supply chain strategy are often – both in theory and practice – "demand" or "supply" characteristics of the supply chain. Let's have a look on some well-known concepts. For example, Christopher develops the following matrix of possible supply chain strategies by specifying "supply" characteristics according to the "lead time of replenishment", and "demand" characteristics by the "demand forecast" of the products (see Fig. 2.4). As a further example, Chopra and Meindl derive the supply chain strategy through specifying "demand" characteristics by the "demand uncertainty" along the whole supply chain (see Fig. 2.5).[7] As a last example, Fisher specifies the "supply" and "demand" characteristics with the following product aspects: the product life cycle, demand predictability, product variety, and market standards for lead times and service. Accordingly, Fisher identifies two categories: products that are either primarily functional, or primarily innovative. Fisher argues that each of these product categories requires a distinctly different kind of supply chain: "Functional products require an efficient process; innovative products, a responsive process."[8] (See Table 2.1).

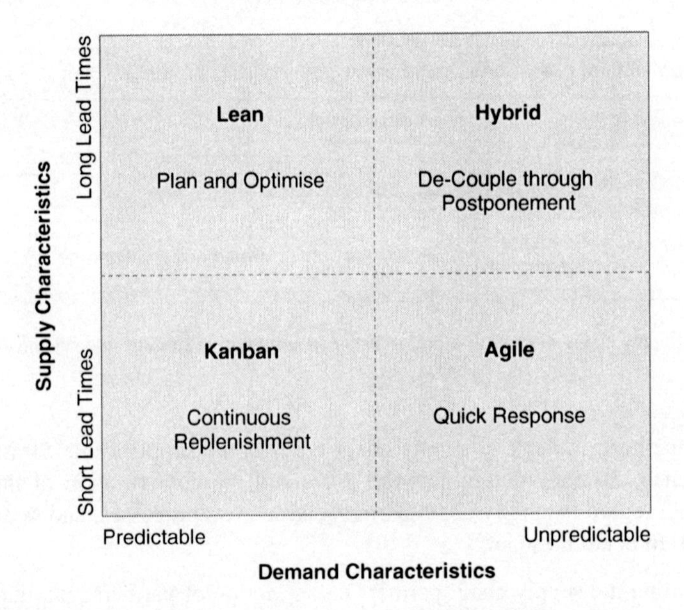

Fig. 2.4 Matrix of generic supply chain strategies (Christopher 2005)

[7] Chopra (2004), p. 3.
[8] Fisher (1997), p. 109.

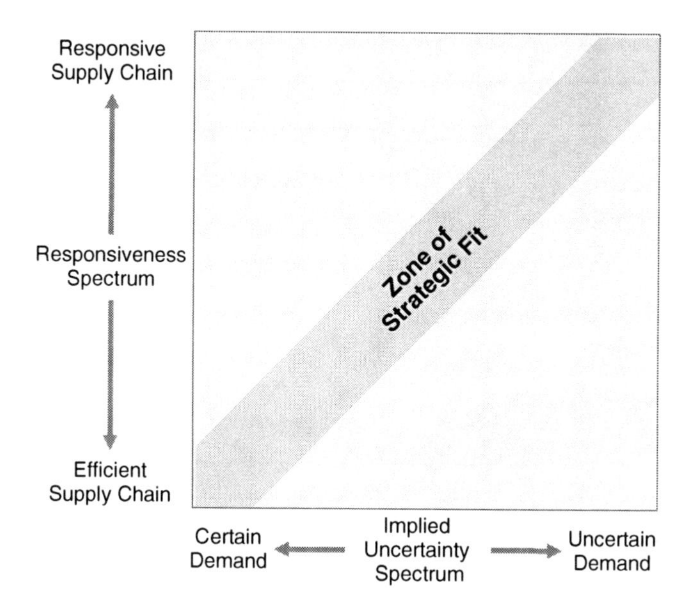

Fig. 2.5 Finding the zone of strategic fit (Chopra 2004)

Having considered these concepts and further discussions with practitioners, bestLog reveals that today's supply chain strategies concentrate more on "supply" and "demand" characteristics rather than on supply chain external conditions, such as social, technological, environmental, and political conditions. And this is a main reason why today's supply chains misjudge the relevance of sustainability in their strategies, although the link to sustainability is immense. As suggested in Fig. 2.3, the type of supply chain and its strategic configuration determine the "volume" and the "quantities" of supply chain operations such as transport, warehousing, production, and recycling. A simple example: a European distribution network consisting of one central warehouse combined with air freight has a different operations volume from a network consisting of seven regional warehouses combined with rail and road freight. Therefore, the social, economical, and energy and resource consumption levels will distinctly differ, in turn influencing efficiency and cost aspects. Further differences will be seen in the CO_2 emissions of the trucks and planes, or in packaging waste in the warehouses.

And this is the main point here. The supply chain, as a key to competitive advantage in many companies, significantly determines the social, economic and environmental impacts of your company, which in turn influence more and more of your stakeholders and shareholders. Hence a *sustainable* supply chain strategy – representing one of the most important success factors for achieving sustainable development for your company–must exhibit the following characteristics:

- It is aligned to the underlying corporate and competitive strategy
- It considers demand, supply, and in particular other, wider general conditions

Table 2.1 Functional versus innovative products: physically efficient versus market-responsive supply chains (Fisher 1997)

	Functional (predictable demand)	Innovative (unpredictable demand)
Aspects of demand		
Product life cycle	More than 2 years	3 months to 1 year
Contribution margin[a]	5–20%	20–60%
Product variety	Low (10–20 variants per category)	High (often millions of variants per category)
Average margin of error in the forecast at the time production is committed	10%	40–100%
Average stockout rate	1–2%	10–40%
Average forced end-of-season markdown as percentage of full price	0%	10–25%
Lead time required for made-to-order products	6 months to 1 year	1 day to 2 weeks

	Physically efficient process	Market-responsive process
Primary purpose	Supply predictable demand efficiently of the lowest possible cost	Respond quickly to unpredictable demand in order to minimize stockouts, forced markdowns, and obsolete inventory
Manufacturing focus	Maintain high average utilization rate	Deploy excess buffer capacity
Inventory strategy	Generate high turns and minimize inventory throughout the chain	Deploy significant buffer stocks of parts or finished goods
Lead-time focus	Shorten lead time as long as it doesn't increase cost	Invest aggressively in ways to reduce lead time
Approach to choosing suppliers	select primarily for cost and quality	Select primarily for speed, flexibility, and quality
Product-design strategy	Maximize performance and minimize cost	Use modular design in order to postpone product differentiation for as long as possible

[a]The contribution margin equals price minus variable cost divided by price and is expressed as a percentage

- It incorporates environmental, social, and economic perspectives in all proposed actions
- It builds increased shareholder and stakeholder value, especially customer satisfaction

A supply chain strategy exhibiting these characteristics is sustainable and represents the starting point of best practice companies to adapt their enablers for managing internals (see Chap. 4), managing externals (see Chap. 5), to adapt their performance measurement system (see Chap. 3), and to adapt their risk management (see Chap. 6). We will now look at the key ingredients of a sustainable supply chain strategy, which we have derived from our bestLog research.

2.3 Ingredients of a Sustainable Supply Chain Strategy

From our countless discussions with hands-on practitioners, industry experts, trade association representatives, academics, and politicians, which we conducted in dedicated workshops, via the bestLog online platform and in case study interviews in the course of the bestLog project, we have identified the distinctive ingredients of a sustainable supply chain strategy – defined in the previous section. The following gives an overview of these ingredients.

2.3.1 Ingredient I: A Strategic and Holistic Approach

The bestLog research shows clearly that today's sustainable practices, balancing environmental, economical, and social goals:

- Are extremely varied and arbitrary
- In large corporations, are often implemented as "island solutions" by individual business units and often with a regional or specific customer scope
- Often lead to solutions which seem to be sustainable at first glance if you look only at the particular company, but not for the total supply chain, where suppliers or service providers are set unbalanced objectives.

These failings, we have found, are due to the absence of:

- Specific sustainability goals in the corporate vision and strategy
- Specific sustainability goals in the supply chain strategy
- End-to-end responsibility of responsible logistics and supply chain managers
- Implementation experience and shared knowledge and
- Top management commitment

Logistics and supply chain managers are often mainly focused on measures that appear to lie within their natural scope of responsibility. They have often implemented environmental and social policies as a kind of aside; often without integrating them with the economic dimension. Sometimes to a greater and sometimes to a lesser degree, they have not really viewed or addressed economic, environmental or social responsibility issues holistically.

Define Sustainability As a Strategic Issue and Be Concrete Best practice companies show different characteristics: Sustainability is a strategic issue and is given top management commitment, with responsibility for sustainability being assigned to management at board level. Sustainability is integrated in the corporate policy and strategy in the form of guiding principles and visions. Based on a top-down approach, sustainability is rooted in the supply chain strategy and sustainability goals are defined in concrete supply chain KPI systems.

Case:
INDITEX Pro-Kyoto Project
INDITEX is developing a Strategic Environmental Plan 2007–2010 at top management level which is divided into five specific projects; one related to logistics.

(See Case Collection in Part IV of this book)

Enhance Learning and Transfer Knowledge from Existing Good and Especially "Worse" Practices Our bestLog research showed that especially large and global corporations lack transparency concerning their designed, planned, and implemented sustainability practices. As mentioned, many practices in such companies are developed by individual business units, and often with a regional or specific customer scope.

In contrast, best practice companies show transparency regarding all designed, planned, and implemented sustainability practices, even along their end-to-end supply chains. They are able to use this information to understand their specific customer and market needs as well as the changing business environment.

Best practice companies do not only focus on successful or best practices. Learnings from failed or "worse" cases can be as valuable as good or best practices, or more so. The bestLog transferability research reveals that sustainable practices have many generic similarities and allow cross-functional learnings, hence can cross organisational classifications and boundaries, e.g. business units, during the knowledge collection and learning phase.

Broaden the Responsibility of Your Supply Chain Managers Supply chain managers in best practice companies are involved in the early stages of product development and product design. Challenges tackled during the product development process rarely concern logistics, on the other hand changes in product development are sometimes very small, but may have a big impact on logistics efficiency and finally, on eco-efficiency. How effective such "simple" things can be is shown in the IKEA case study.

BestLog research showed that best practice companies emphasise intra-organisational communication and enlarge the sphere of responsibility of their supply chain managers.

Case:
IKEA – Air Hunting Competition
It is interesting to see that a single person at IKEA is entirely responsible for the Tealight product supply chain. The so-called "Need Planner" has full transparency along the whole end-to-end supply chain. This enabled this individual to explore new efficiency potentials regarding the product's packaging and "simple" product re-design.

(See Case Collection in Part IV of this book)

Establish a Dedicated Organisation, Train and Motivate People New sustainability initiatives and projects along the supply chain may have a broad impact on several operations, within and beyond individual companies. This may cause additional complexity and more intra- and inter-company trade-offs, especially due to knowledge being dispersed along the supply chain. Our research shows that in consequence some large best practice companies have extended the scope of their existing health, safety and environmental protection (HSE) organisations to encompass supply chain management issues. The role of such a separate, centralised, cross-functional, organisation for sustainability is to:

(a) Report to top management
(b) Consolidate issues and support implementation projects
(c) Provide advice
(d) Collect and disseminate learning and best practice
(e) Represent a single point of contact for external entities and supply chain partners
(f) Organise and coordinate training
(g) Consolidate R&D activities

Another role of this organisation is to train personnel and to develop tools and instruments for their day-to-day, as the implementation process for sustainability goals is the key to winning acceptance and achieving *durable* sustainability solutions. The effectiveness of training and implementation tools can be enhanced by an incentive system. Incentives behind sustainable solutions can be monetary, but may also recognize achievements with awards or certificates. Monetary rewards can be linked to cost savings.

Sustainability is an Opportunity. Invest in R&D Depending on your industry's structure, you may need to push for R&D activities with regard to sustainability. One reason may be a business's high dependency on natural resources; this applies in particular to Logistics Service Providers (LSP). Natural resources are becoming increasingly scarce and expensive. This, together with the massive demographic changes afoot worldwide (population growth, urbanization) leads to huge opportunities, but also serious risks.

The bestLog research shows that best practice companies integrate sustainability with their R&D activities, focus on broader challenges, are on the whole optimistic, and see environmental issues as providing an opportunity to reduce costs in the supply chain and to open up new markets by developing environment-friendly products and services.

Create Values, Knowledge, and Culture with the Right People and the Right Image The bestLog research showed that good concepts behind best practice have often been developed from simple ideas by the people directly responsible. The key lever to implement a sustainable supply chain with a common vision and culture is the people, the actors who make the decisions in the supply chain. The values, awareness, and the mindsets of the people in your supply chain are fundamental in determining your sustainability results. Development begins with the selection of your supply chain partners and your employees.

In this context, the bestLog research shows that best practice companies put strong emphasis on sustainability at executive and senior management level, as sustainability values are generally transferred top-down to employees.

Another, associated finding was that best practice companies merge sustainability with their general company image. One reason: employees tend to work in line with the overall corporate image, which is communicated internally as well as externally. A good example is "Quality Thinking" at Mercedes-Benz, a slogan which is strongly communicated by the company since its foundation, and firmly anchored in the behaviour of its employees.

My belief is that a strong corporate culture with an explicit value system will (in the long run) lead employees who do not follow these values to leave the company. And those who do voluntarily espouse the corporate values will transfer these values along the supply chain, and will assess their supply chain partners based on this value system. A strong sustainability culture at company level will be transferred across the whole supply chain – assuming the company concerned is the dominant partner in the supply chain.

The bestLog research also indicates that best practice companies emphasise the creation of sustainability knowledge and the implementation of practical tools. Training and education play key roles. Focus on specific training topics is used to establish the value system. The organisation's vision and its values are incorporated within concrete competencies, actions, and instruments.

2.3.2 *Ingredient II: Recognizing and Assessing Current and Future Trends*

Over-arching macroeconomic and social trends,[9] globalisation and increasing competition, longer and more fragmented supply chains, and increasing stakeholder and shareholder demands, all affect the complexity and dynamics of supply chains. More than ever, today's supply chain managers are confronted with dynamic and discontinuous change (e.g. oil prices), and the more dynamic they are, the harder trends are to forecast. Longer and increasingly fragmented value chains, together with totally new dimensions of objectives (e.g. CO_2 emissions reduction), extend the responsibilities and agendas of today's supply chain managers.

Future-oriented sustainable strategies must be able to adapt to nascent trends as early as possible, and logistics goals must be geared towards these trends. This is

[9]Changes in the business environment occur usually as trends. Trends in a business environment are changes which take place over time and affect companies in their competitive environment; for example, the current "corporate social responsibility" trend or the "green SCM" trend. Some such trends – so-called "mega trends" – have a more global and extended impact on the economic, business, and social environment. Climate change is an example of such a mega trend: It affects customers, regulations, society, the competition, investors, and the markets of a company.

key to ensuring the long-term success of a company. The challenges for supply chain managers in this context are (a) to identify the trends which are relevant to their complete supply chain and (b) to assess and evaluate their potential negative and positive impacts.

"Scenario-Planning": Bring Supply Chain Managers and Stakeholder Executives into the Process Early "Scenario planning" is a traditional tool to plan and develop different paths (=sequences of different scenarios) for your business with regard to trends. Scenario planning is now a standard tool in companies and is often applied successfully, due to its simplicity. The bestLog research shows that today's uncertainties complicate scenario planning efforts: the number of variables at play – and the range of plausible outcomes – have exploded in recent years, especially in the context of global supply chains. Different outcomes for each of the supply chain uncertainties may produce very different paths for the business.

Best practices companies manage this challenge differently: their supply chain strategists apply a demanding process of gathering information and exploring possibilities. Best practice companies bring supply chain managers along the supply chain and stakeholder representatives into the planning process early. This inclusive approach inculcates an appreciation of the threats the supply chain faces and leads to a collective strategic response.

Be Aware of Uncertainty Regarding Sustainability Issues in Your Supply Chain The bestLog research shows that best practice companies recognise trends in their business environment earlier than their competitors, and are better able to assess their possible impact. Hence, best practice companies embark on critical action, requiring long lead times, earlier; e.g. early investments in given technologies, the development of strategic partnerships, or the development of innovative products or services.

Furthermore, best practice companies monitor trends in a systematic and holistic manner, using indicator systems to observe and analyse all potential developments in their business environment. Best practice companies are able to solve the associated technical, structural, and organisational issues, a capacity especially relevant to global supply chains. And again, a key success factor in their supply chains is their broad, end-to-end responsibilities.

Understand Cause–Effect Relations Between Trends and Your Supply Chain The early identification of trends is one thing: the next challenge is to assess the potential impacts of those trends on your supply chain. Here are some examples of quite difficult questions which best practice companies were able to answer in our interviews: What is the total cost effect for your supply chain if the oil price doubles? What oil price justifies structural change in your supply chain, e.g. regionalisation? How far could the CO_2 emissions of your supply chain be reduced compared to your company-wide CO_2 emissions? How much has your practice reduced your external costs? We have seen that best practice companies try hard to understand the complex cause-and-effect relations between uncertain trends and their supply chains.

Case:
Mercadona and Renfe Collaboration on Intermodal Distribution
Mercadona has achieved an external costs saving of € 13.1 million.
(See Case Collection in Part IV of this book)

2.3.3 *Ingredient III: Implementation of Economic, Environmental, and Social Objectives*

Implementation – integrating objectives into operations – plays an important role in the adaptation of supply chains to "new" supply chain objectives such as environmental or social goals. Implementation is, put simply, the "translation" of strategic, partly generic, long-term goals into operational, specific, short-term goals. One result of an implementation process is usually a group of interrelated KPIs, also called a "KPI system". Such a KPI system is a kind of "quantitative tool" used to implement strategic goals into company operations. Of course there are, besides a quantitative "tool", also pretty important qualitative aspects influencing a strategy implementation, e.g. good communication, high motivation, the right awareness, etc. We will resume these aspects in the next section.

The challenge for supply chain managers is to "break down" the strategic goals into the "right" KPIs and to define the specific target figures for a given reporting period. The challenge in selecting the "right" KPIs is, of course, to select those KPIs which measure and control the right operational processes. Another challenge is that different operational goals may be complementary, or may conflict with each other; a KPI system must take potential trade-offs into account.

Environmental and social goals in particular are in most cases new to many operations, and need to be integrated into their existing KPI system. One recent study on trends and strategies in logistics showed that 43% of large companies and only 26% of SMEs had defined concrete environmental and resource protection goals among their logistics operations' targets.[10] One reason is a lack of knowledge regarding the measurement and the assessment of environmental and social impact KPIs.

Develop KPI Systems Taking into Account New Kinds of Goal and Related Trade-Offs The bestLog research shows that best practice companies tend to have the capability to translate new strategic goals[11] into the operations via an adaptive and intelligent KPI system. Such KPI systems consider trade-offs and complementarities between social, environmental, and economic goals.[12] One success factor is the strong top management commitment to these goals. Another

[10]Straube and Pfohl (2008), p. 69.

[11]New also in terms of the dimension considered; social and environmental goals within a supply chain strategy are in many companies new, or at least secondary.

[12]We will return to the topic of "measurement" and KPI systems again in Chap. 3.

success factor is that the supply chain managers of best practice companies are able to describe cause-and-effect relations in their supply chain with regard to social, economic, and environmental issues.

2.3.4 Cash Flow Matters: Link Social and Ecologic Goals with Financial Figures

A green strategy or a social strategy is not a strategy for sustainability. True sustainability must give equal weight to the economic dimension. Sustainable supply chain practices must be financed and provide pay back within a reasonable time span. In bestLog we have seen logistics practices which were abandoned during the economic crisis in 2008 due to their economic disadvantages. An example was Mercadona's intermodal transport practice in Spain, which caused clear negative cash flow, having not reached critical transport mass.

The challenge is to assess the economic, in particular financial, aspect of a practice. Best practice companies are capable of this. They have extended their supply chain KPI systems to include social and environmental measures and can link this whole operational KPI system with their financial measurement systems. The Sustainable Supply Chain (SSC-)Scorecard in Sect. 2.4 will provide a generic template to apply this approach in your company.

2.3.5 Ingredient IV: Balancing Economic, Environmental, and Social Objectives

Part I has shown the general relevance and importance of balancing social, environmental and economic objectives in companies' sustainable development. The central challenge here is that balancing economic, environmental and social objectives requires first of all an understanding of the conflicting and complementary relationships between them. The strategic challenge is than to establish realistic and balanced targets in different operations in all three dimensions. And each of the three dimensions must be aligned to the overall corporate strategic goals and vision, regardless of possible specific trade-offs.

Provide Incentives and Motivate People The bestLog research shows that best practice companies fulfil a necessary condition of implementing balanced goals in an organisation: they provide incentives and motivate the people involved in order to change their attitudes and to redirect their business systems towards these balanced goals.

So a sustainable supply chain strategy must rethink existing intra- and inter-company incentive systems in order to tackle unbalanced objectives. An important strategic question which we discussed several times in bestLog was: Do the current business models applied by supply chain partners motivate and reward a balance of sustainability goals? The answer was quite clear: not in the majority; supply chain managers are usually driven by traditional cost- and service related measures. Achieving social and environmental objectives, although they might be complementary to other objectives, is in most cases not directly rewarded.

Case: IKEA – Air Hunting Competition
The interesting learning of this case is that IKEA started an internal competition motivating all people to find highest amount of air in their transport processes. The competition itself was a highly motivating factor.

(See Case Collection in Part IV of this book)

Make People Aware of the Long-Term Benefits Further, the bestLog research revealed that making people aware of the long-term benefits of change and implementing best practice into day-to-day work is an important strategic challenge, addressing different strategic management levers such as branding, communication, culture, knowledge management, knowledge development, trust and collaboration, as well as the transparency of cause-and-effect relations.

The next section takes these challenges and ingredients of success, and describes a general, iterative approach with a number of tools and frameworks to support you in the development of your sustainable supply chain strategy.

2.4 An Iterative Approach to Developing Your Sustainable Supply Chain Strategy

The ingredients of a sustainable strategy were identified in our bestLog project research. The following describes briefly an iterative six-step process approach to integrating these principles into your existing supply chain.

A sustainable supply chain involves more than the implementation of popular practices – most of them are just building blocks. Rather, individual practices must be assembled to integrate meaningful long-term sustainability principles, along the end-to-end supply chain. A systematic approach to strategy design and integration can help companies developing a sustainable supply chain to create a value proposition.

The following five questions define the road map to a sustainable supply chain (see Fig. 2.6), helping to change or redesign your current supply chain strategy. The action plan will be derived from the risks and opportunities your supply chain faces in the present business environment, and will face in future. The approach ends with a Sustainable Supply Chain (SSC-)Scorecard, which serves aid in the implementation of a strategy for sustainability into your existing supply chain.

This iterative six-step approach has to be seen as a cycle which should be executed regularly in your supply chain, since relevant conditions may change quite quickly and sometimes radically – see, for example, the oil price spike of 2008.

Step 1 aims to take stock of the current state of company- and supply chain-specific characteristics regarding strategy, resources, and current and planned practices. It is mainly concerned with internal factors and considers elements which are usually within the control of a company. Step 2 aims to identify current and forecast potential future developments and trends, focusing on external factors influencing the supply chain. It considers factors which are usually not under the direct influence of a company. Step 3 aims to evaluate the risks and opportunities derived from these internal and external factors. It serves to define company- and supply chain-specific "sensitivity". Step 4 takes this analysis to the existing supply chain strategy, and institutes a strategy change or redesign process with regard to the sensitivity identified. Step 5 focuses on implementation issues in order to balance social, economic, and environmental objectives, with the aid of a novel Sustainable Supply Chain Scorecard concept. Step 6 focuses on the key ingredients required to successfully implement the sustainable supply chain strategy in the relevant organisations.

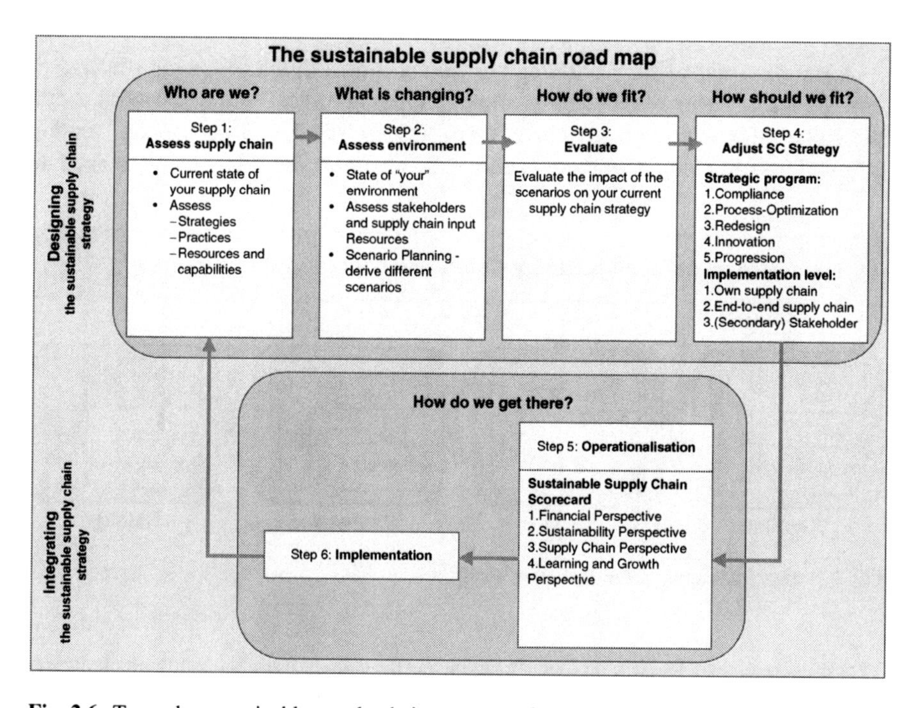

Fig. 2.6 Towards a sustainable supply chain strategy – from principles to practice

2.4.1 Step 1. Stocktaking: The Current State of Your Supply Chain

The "as-is" analysis in this step 1 is focused on the strategies, the sustainability practices and the associated resources and capabilities in your supply chain. The assessment of these attributes will help later in step 3 to identify the risks and opportunities you are facing, and to derive the appropriate extension or re-design of your existing supply chain strategy in step 4.

Strategies. Aiming to match the strategic and holistic approach of best practice companies (see Sect. 2.3.1) requires an understanding of the existing strategies and the associated strategic goals within your own company and along your supply chain. The integration of a sustainable supply chain strategy is usually not a greenfield development, and will most likely not lead to a complete re-engineering of your supply chain and strategy. Rather, step-by-step adaptation is preferable, unless your company and its supply chain are urgently threatened by the changing business environment.

So, take stock of:

- The existing corporate and competitive strategies
- The sustainability strategies (if sustainability is not already part of the corporate strategy)
- The company-specific supply chain strategies, and finally
- The cross-company supply chain and collaboration strategy[13].

A simple result of such a strategy review is shown in the following illustration (Fig. 2.7). Examples of those basic strategy types are listed in the following table. We will not focus here on any one of them – please see the "Bibliography" section for more information about them (Table 2.2). The most important result of such an

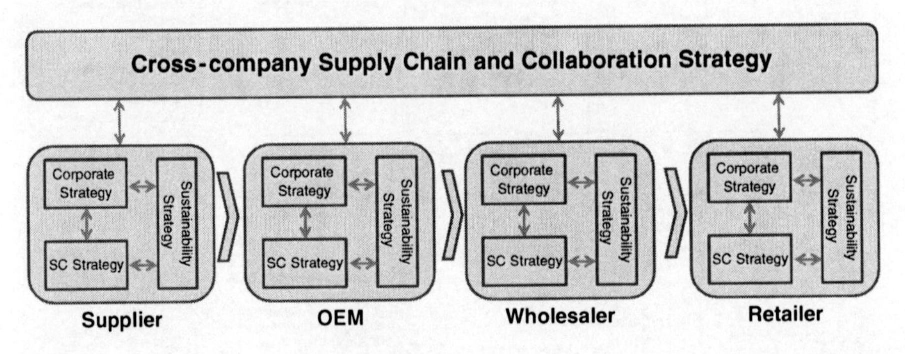

Fig. 2.7 Strategies along a simple supply chain

[13]For the sake of completeness we should mention that the company-specific supply chain strategy of many OEMs determines substantially the cross-company supply chain strategy of the whole supply chain.

Table 2.2 Strategies along a simple supply chain

Strategy level	Strategy type	Strategy type
Competitive strategy	Cost leader	Differentiation strategy
Supply chain strategy	Lean, cost- and efficiency driven	Agile, service- and speed-driven
Cross-company collaboration strategy	Hierarchal control	Heterarchal control
	Long-term partnership	Short-term partnership
	Horizontally integrated networks	Vertically integrated networks
Company's own sustainability strategy	Defensive	Offensive

analysis of the *status quo* is the understanding of potential intra- and inter-organisational goal conflicts. It seeks to discover whether there is a common strategic alignment, and whether the goals of each strategy element are integrated, aligned, and complementary. A company's own supply chain strategy which is not aligned to the corporate and cross-company supply chain strategy cannot per se be sustainable.

Potential short- and long-term goal conflicts can rapidly become serious barriers to implementing a sustainable supply chain along its members; especially if the supply chain is required to adapt quickly to change.

The analysis will also show if top management commitment is given within the individual companies and the supply chain. The section on the Ingredients of a Sustainable Supply Chain Strategy showed that the lack of top management commitment is the most common cause of "island" solutions.

A holistic understanding of the existing business models and their goal relations in such a system will also allow a better balancing of economic, environmental and social objectives, which may also result in a change of existing business models between supply chain partners.

Practices, Resources, and Capabilities. With regard to best practice companies your stocktaking analysis should also serve to increase transparency, revealing existing, planned, and failed sustainability practices; particularly along the end-to-end supply chain. This allows you to understand on the one hand the trends in your supply chain regarding your customers' requirements and the market's needs, and on the other, to see where skills and resources may be lacking.

Step one ends with an "as-is" analysis of the current state of company- and supply chain-specific characteristics. The next step aims to identify current and forecast potential future developments and trends emerging from external factors influencing the supply chain. It is an external view, and considers aspects which are usually not under the direct influence of a company.

2.4.2 Step 2. Your Environment: Current, Potential, and Future Impact Factors

The second step in our approach is linked to Ingredient II and deals primarily with what is changing in the business environment, what kind of scenarios your company will face in the medium and long term, and finally, what the main driver of change may be. This will lead, together with the "as-is" analysis from step 1, to the definition of potential risks and opportunities for your supply chain strategy in step 3.

The most important topics to be considered in this step are supply chain input resources, such as fuel and energy, stakeholders, and shareholders.[14] We have devoted a separate chapter to the topic of supply chain stakeholder management, in recognition of its relevance to sustainable supply chain management.[15] The key task – details will be discussed in this later chapter – is to identify the "right" stakeholders and their "stakes" in your supply chain in order to get a precise assessment of their risk and opportunity factors for your supply chain – now and in future. The potential stakeholders in a supply chain are shown on the following illustration (Fig. 2.8).

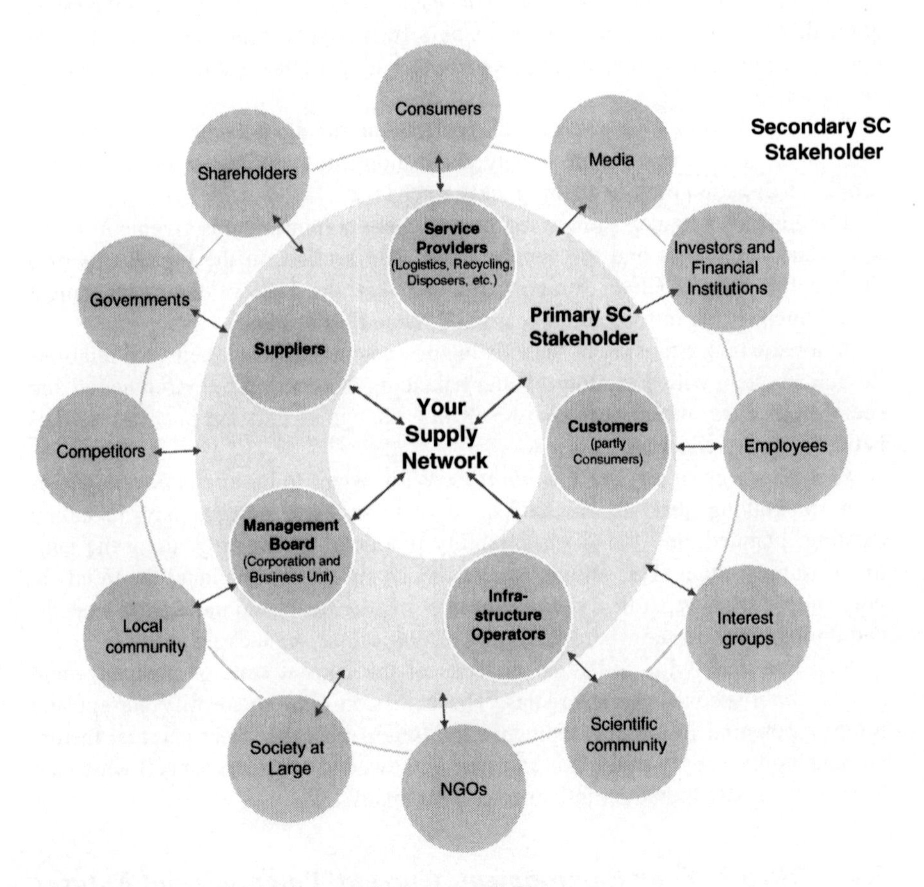

Fig. 2.8 Primary and secondary supply chain stakeholders

[14]Some stakeholder theories consider shareholders also as stakeholders. I do so, too. See Chap. 6.
[15]See Chap. 5.

Supply chain input resources such as fuel, energy, and natural resources nowadays deserve close attention in supply chain management logistics. The trend of rising prices and increasing scarcity make input resources major risk management factors in an economic perspective, especially if you run cost- and energy-sensitive supply chains, such as the commodity micro chip industry with its international production and transport flows for example. Understanding and forecasting input resource-related information helps in developing your sustainable supply chain strategy.

One well known and useful tool to use in step 2 is "Scenario Planning". This is a method of medium- and long-term planning, simulation, and of forecasting probable future developments based on the continuous observation of indicators. Scenarios allow supply chain managers to gain a better understanding of the possible business environments they will need to tackle in the future.

The first step in determining the scenarios is to explore the drivers that are most likely to shape the future of your supply chain. These are primarily, as mentioned earlier, the stakeholders and the supply chain input resources. For practical reasons, the number of possible basic scenarios is usually limited to the following three types: optimistic, pessimistic, and most likely. The analysis of existing practices during step 1 may help to determine some of the drivers and trends. In a highly uncertain business environment as described earlier, it makes sense to analyse the drivers according to two criteria: first, their predictability, and second, their impact on your supply chain (see Fig. 2.9).

Step 2 ends with a set of current, potential, and future requirements for different stakeholders and input resources. These feed into the next step, where they will be assessed together with the status quo of the supply chain.

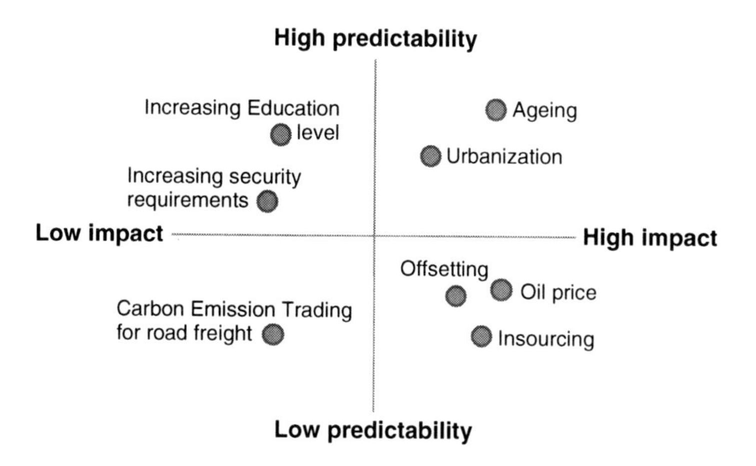

Fig. 2.9 Clustering scenario drivers – examples

2.4.3 Step 3. Evaluation: Identifying Potential Risks and Opportunities

Based on the assessment of your supply chain and of your business environment, you can now identify potential risks and opportunities. These will serve in step 4 to change or re-design your existing supply chain strategy. We have given a separate chapter to the topic of risk management, as increasing complexity, accelerating change and uncertainty are posing growing challenges for supply chain managers on the way to a sustainable supply chain.[16]

The following evaluation framework allows you to assess your supply chain capabilities in the context of the scenarios you have prioritised from step 2. The assessment then serves to determine your supply chain specific Threat-Opportunity-Profile (see Fig. 2.10). You need to understand the cause-and-effect relationships between potential success factors to undertake this evaluation. For example, you should be able to estimate that the regionalization of procurement structures in response to an oil price increase would be likely to reduce your transport costs by *x* %. Only an effective understanding of the relevant levers in your supply chain and of their potential impact on the planning scenarios permit precise analysis of strengths and weaknesses and finally the "right" definition of risks and opportunities.

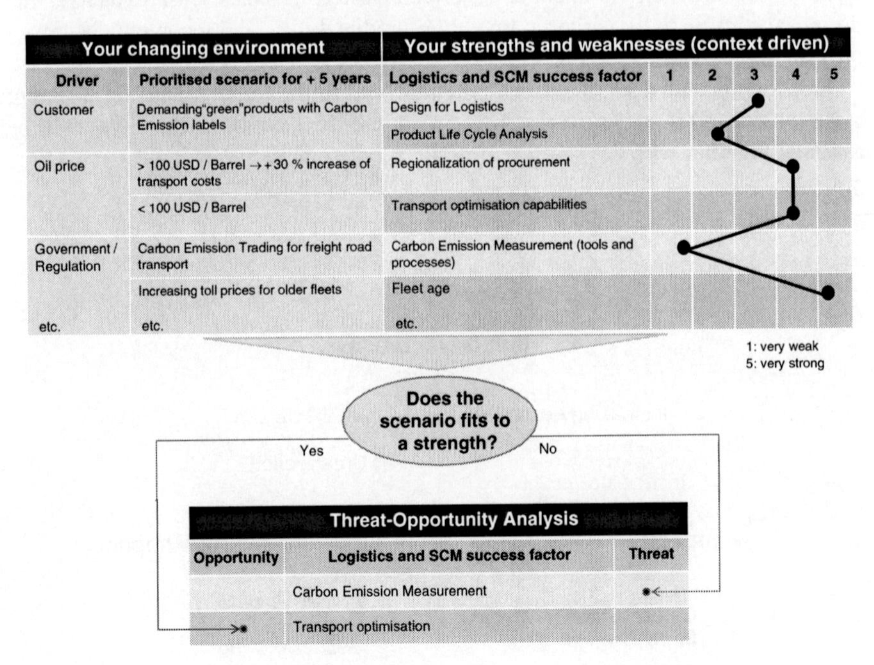

Fig. 2.10 Threat-Opportunity-Profiles for the sustainable supply chain strategy

[16]See Chap. 5.

2.4.4 Step 4. Action Plan: Extend or Re-design the Supply Chain Strategy

Based on the Threat-Opportunity-Profile, you can now define strategic gaps on the route to a sustainable supply chain; gaps between your current supply chain strategy and the changing business environment. The greater and more relevant the potential risks and opportunities, the larger will be the gaps, and the greater the need to act and make strategy changes. As mentioned earlier, the integration of sustainability principles into the supply chain is usually not a greenfield development; it has to be a step-by-step approach.

The scope of this supply chain review can vary. The larger the gaps, the broader the implications for implementing change, and the more aggressive the strategic programme and the associated action plan must be. The extension or the re-design of your existing supply chain strategy should be precisely defined. Of course, this is a strongly context-driven process, and based primarily on the results you have now obtained from this six-step approach.

Let's have a detailed look at following Fig. 2.11. The *implementation level* determines who you involve in the implementation of initiatives. These actors could be the members of your own supply chain, which you influence directly; they could extend further to scope 1 externals from the end-to-end supply chain (primary supply chain stakeholders), or yet further to scope 1 and 2 (secondary) stakeholders. The implementation level involves more than the parties involved; it is also defined by the time horizon you consider in your program.

The implementation level should not be mixed up with the driving forces you take into account when developing your plans. For example, the "compliance" program naturally requires a good understanding of governmental and regulatory

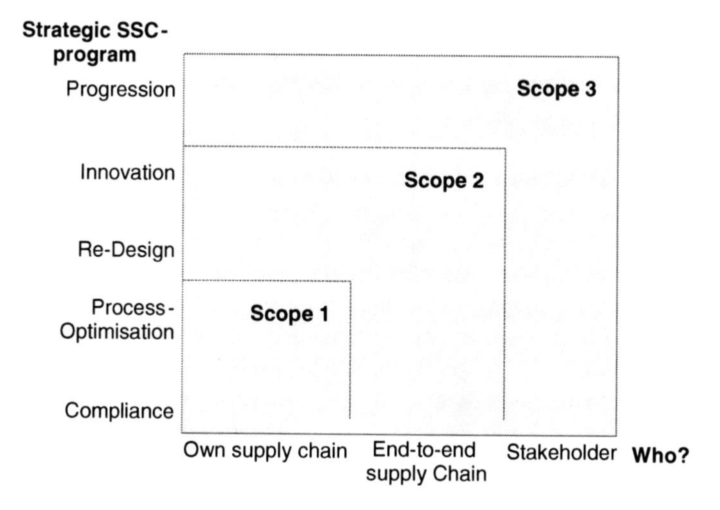

Fig. 2.11 Scope of sustainable supply chain strategy implementation

requirements, but you will not aim to develop and implement compliance solutions proactively together with policy makers.

The *strategic sustainable supply chain program* (*SSC-program*) determines the set of actions/initiatives needed to close the gap and to create your customized sustainable supply chain on the long-term; customized because it fits your gap precisely. The following strategic programs can be distinguished:

1. Compliance
2. Process-Re-engineering
3. Restructuring
4. Innovation
5. Progression

Again, the strategic program does not represent a new supply chain strategy, rather a program to extend or re-design the existing strategy. And again, the bigger the gap, the stronger and more aggressive will be the action that needs to be taken. The first two programs are essentially defensive and focus mainly on the company's own supply chain. The last three programs are more aggressive and require an extended implementation level beyond the company's own supply chain. Below we will take a brief look at each of these strategic programs, together with some examples and case study references:

Strategic SSC-Program: "Compliance" Supply chain managers should follow the "Compliance" program to obtain the benefit and competitive value of "Reducing and managing risk", which represents a precautionary response to stakeholders, shareholders, owners, and employees in particular. This program is mainly driven by the compliance requirements of stakeholders such as governments, customers, and special interest groups. Supply chain managers satisfy their stakeholders' concrete requirements and monitor developments in the regulatory arena. Some examples of program actions, described in part in the bestLog case studies, include:

- Certification of adherence to mandatory standards
- Certification of adherence to customer-specific standards
- Monitoring of planned new regulations

Strategic SSC-Program: "Process Optimisation" Supply chain managers should follow the "Process Optimisation" program to obtain the benefit and competitive value of "improving productivity and efficiency" and in consequence, of reduced supply chain costs, increased resource productivity, and reduced environmental impacts. This program is mainly driven by additional supply chain and logistics costs, due to e.g. increasing resource prices or increasing regulation compliance costs.

Actions taken to improve efficiency do not aim to change existing structures; rather try to improve the existing system itself, based on optimisation and improved planning. The overall goal is to invest in improvements showing a positive net present value in the short term; this is difficult in many cases of current, existing "green" technologies, for example. The scope of such activities starts within the company and should be extended to the whole supply chain. The first activities should focus on the "low-hanging fruit" and on the areas in which supply chain managers already operate and respond to change, e.g. transport planning. Some examples of program actions, some of them described in the bestLog case studies, include:

- Training of truck drivers
- Simple re-design of packaging or optimised packaging processes
- Implementation of environmental management systems, to standardise processes and to reduce unnecessary complexity costs.
- Consolidation of material flows by re-scheduling
- Route optimisation and transport planning
- Business process optimisation to reduce lead times, which enable longer planning cycles, which in turn create more consolidation possibilities
- Applying intermodal transport
- Applying fleet management systems
- Energy management in warehouses

Shift from Defensive to Aggressive Strategic SSC-Programs These first two strategic SSC-programs fit very well for companies and supply chains with cost leadership strategies and efficient supply chain processes. Successful implementation of these programs can satisfy competitive requirements, if the regulatory and cost risks – defined by the Threat-Opportunity-Profile – remain only potential and do not affect long-term competitiveness.

But one common finding in the literature and within the bestLog practitioner community is that cost advantages based on efficiency and productivity concepts do not lead to a long-term, sustainable competitive advantage; this is because (a) competitors will most probably have the same opportunities to emulate such cost advantages and (b) efficiency and productivity improvements fail, if the concepts do not achieve the required critical mass of throughput: for example, during the 2008 global economic crisis. And in addition to that: The high investment demands of logistics-related technologies (e.g. regenerative energy, alternative-fuel engines, telematics, etc.), have often long payback periods and lead to higher logistics costs, which such companies will most likely pass to the customers or partners, if they stick to the mentioned cost leader strategy. This means that cost leader companies with cost-efficient supply chain strategies face big challenges in aligning efficiency strategies with sustainability principles in the long run.

In conclusion, the stand-alone integration of the "Process-Optimisation" and "Compliance" SSC-programs will not necessarily lead to competitive advantages, if the strategic gap in the context under consideration is large and very diverse, and along the main supply chain; in more concrete terms, you should shift to more aggressive strategies and extend programs to further implementation levels when stakeholders create strong direct and indirect pressure on the supply chain, or when sustainability drivers such as, for example, rising resource prices impact forcefully on current and future supply chain performance.

Strategic SSC-Program: "Re-design" Supply chain managers should follow the "Re-design" program to obtain the benefit and competitive value of "the long term and cardinal improvement of effectiveness and early prevention of risk". This program is mainly driven by additional supply chain and logistics costs resulting from rapidly increasing resource prices and regulatory costs which affect a company's competitive position. The actions around "Re-design" aim to change existing structures and processes. The scope of activities covers the whole supply chain. Hence, a collaborative approach is one key success factor. The decision making processes needs to be quantified as much as possible, in order to draw up and assess several scenarios, since re-design actions are mostly linked to high costs and investments, and are often irreversible. Some examples of program actions, some of them described in the bestLog case studies include:

- Regionalization of procurement and production structures
- Hub strategies in the distribution network
- Closed-loop supply chain management
- Re-assessment and substitution of suppliers and logistics service providers
- Business Process Reengineering (BPR)

Strategic SSC-Program: "Innovation" Supply chain managers should follow the "Innovation" program to obtain the benefit and competitive value of "differentiation". This program is mainly driven by external stakeholders: customers, consumers, NGOs and suppliers who demand new solutions, products, and services.

The actions around "Innovation" aim to change existing business models, to break existing mindsets in the supply chain, and to achieve a sustainable[17] image, which, in the end, will increase the credibility of a company. The scope of activities is mainly intra-organisational, e.g. between R&D, SCM, manufacturing, and sales as well as inter-organisational in the end-to-end supply chain. Hence, communication and awareness are key success factors; top management commitment throughout the key supply chain players is a must. Some examples of program actions, some described in the bestLog case studies, include:

- Innovation management for new products and supply chain services
- Price differentiation for premium, sustainable supply chain services and products
- Design for logistics (product and packaging)
- Training and education
- Incentive systems for employees and partners
- Profit-sharing models for supplier and logistics service provider
- Remanufacturing
- Knowledge and best practice platforms
- Carbon footprint labels on products and services

Strategic SSC-Program: "Progression" Supply chain managers should follow the "Progression" strategy to obtain the benefit and competitive value of

[17]Not just strong communication and marketing, rather a structural internal change.

"differentiation, first mover advantage, and establishing market entry barriers". The main drivers in the relevant industry, and even society as a whole are: a lack of standards, of knowledge, and of regulations, highly developed corporate social responsibility, and a lack of common market direction.

The actions around "Progression" do not have a direct, measurable payback for a given time period. The benefits are more long-term and qualitative, and go along with reputation and image. The reach of activities is wide spread mainly found in the relevant sector, at governmental institutions, at associations, and in different countries. Hence, a focused approach with concrete milestones is a key success factors, top-management commitment along the key supply chain players is a must. Some examples of actions, some described in the bestLog case studies, include:

- Investing money and time in establishing new standards in the market, e.g. a CO_2 emission measurement standard in the transport sector.
- Funding research institutions and research projects and surveys.
- Funding of relevant NGOs
- Active involvement in associations
- Establishing knowledge exchange platforms
- Communication, discussion, and consulting with politics and regulation.

2.4.5 Step 5. Implementation with the Sustainable Supply Chain (SSC-)Scorecard

What is the status of our iterative six-step approach with respect to the identified main ingredients of a sustainable supply chain strategy? Ideally, we will have selected the appropriate scope of strategic vision and determined the right related actions to align your supply chain strategy to the changing business environment. Most future trends and scenarios have been considered and taken into account. Finally, these findings have been incorporated into the reformulation or re-design of your existing supply chain strategy and the associated objectives. The next question is, "How to get there?"

We are still facing the challenges of making strategy ready to put into operation, and of balancing economic, environmental and social objectives. Potential trade-offs along the supply chain are still not visible. The lack of cause-and-effect understanding is likely to lead to an imbalance in these three dimensions of sustainability. The capability to translate the new strategy elements into a structured KPI system, explaining cause-and-effect relationships and justifying implemented practices, remains a challenge ahead of us. Step 5 will address these issues.

One well known tool/concept for translating strategic goals into operations is the "Balanced Scorecard" developed by Kaplan and Norton from 1990. The Balanced Scorecard is a (performance) management system providing a framework to translate a strategy into balanced operational terms via objectives and measures,

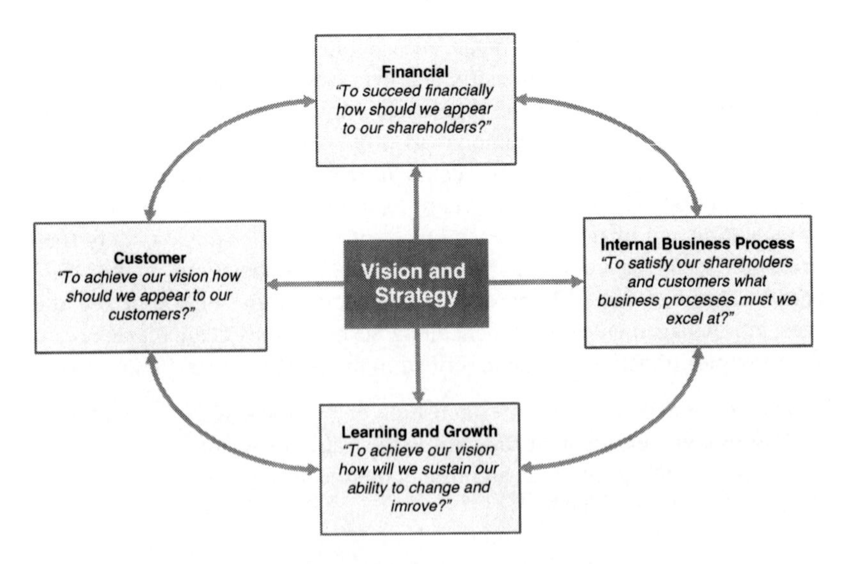

Fig. 2.12 The Kaplan/Norton Balanced Scorecard

organised into four different perspectives: financial, customer, internal business process, and learning and growth (see Fig. 2.12).

"The Balanced Scorecard expands the set of business unit objectives beyond summary financial measures. Corporate executives can measure how their business units create value for current and future customers and how they must enhance internal capabilities and the investment in people, systems, and procedures necessary to improve future performance."[18]

The measures represent a balance (Kaplan and Norton (1996), p. 9)

- Between *external* measures for shareholders and customers, and *internal* measures of critical business processes, innovation , and learning and growth,
- Between the *outcome* measures – the result from past efforts- and the measures that *drive future* performance, and
- Between *objective*, easily quantified outcome measures and *subjective*, somewhat judgmental, performance drivers of the outcome measures.

These characteristics fit ideally with our ingredients of a sustainable strategy and justify the use of this general concept, but not as a pure performance measurement system. "Many people think of measurement as a tool to control behaviour and to evaluate past performance. The measures on a Balanced Scorecard are used in a different way – to articulate the strategy of the business, to communicate the strategy of the business, and to help align individual, organisational, and cross-departmental initiatives to achieve a common goal. Used in this way, the scorecard does not strive to keep individuals and organisational units in compliance with a pre-established

[18]Kaplan and Norton (1996), p. 8.

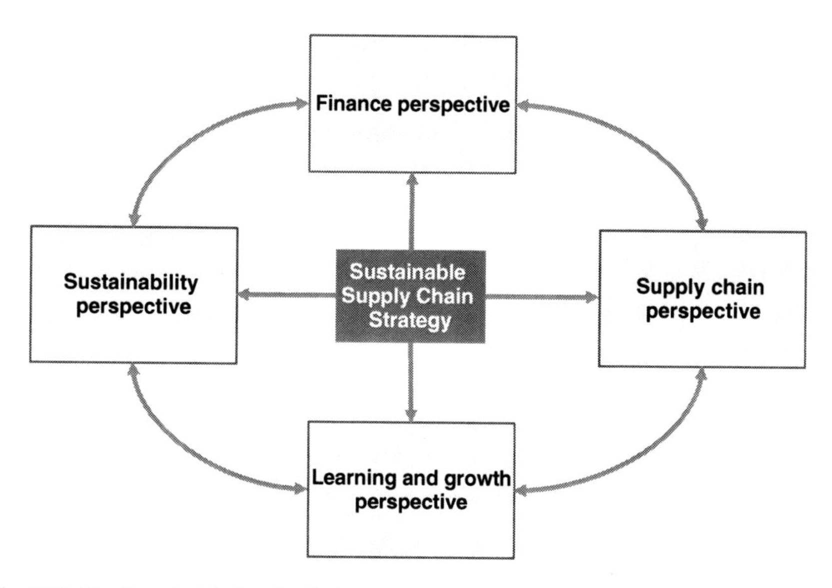

Fig. 2.13 The Sustainable Supply Chain (SSC-)Scorecard Perspectives

plan, the traditional control system objective. The Balanced Scorecard is used as a communication, informing, and learning system, not a controlling system."[19]

The Balanced Scorecard is in use in many companies worldwide. Its success led to the application of the concept to other areas such as marketing, environmental management, production – or supply chain management. Indeed, several sustainability scorecards and supply chain management scorecards have been developed, but separately. Literature analysis on well known scorecards shows that none of them integrate both perspectives. The following illustrations show such an integrated Sustainable Supply Chain (SSC-)Scorecard and the relations between its four perspectives (Fig.2.13 and 2.14). As sustainability basis on and implies a "balance", the word "balanced" was taken out of the designation.

The comparison of the SSC-Scorecard with the "traditional" Balanced Scorecard of Kaplan and Norton shows the structural similarities. The four perspectives, Finance, Customers, Processes, and Learning and Growth, are retained, and are extended by several other aspects regarding supply chains and sustainability. The main reason to stick with the traditional four perspectives is to allow an "easier" integration and "plug-in" to a company's existing used of a "traditional" Balanced Scorecard. Let's have a detailed look on the SSC-Scorecard on Fig. 2.14.

[19]Kaplan and Norton (1996), p. 25.

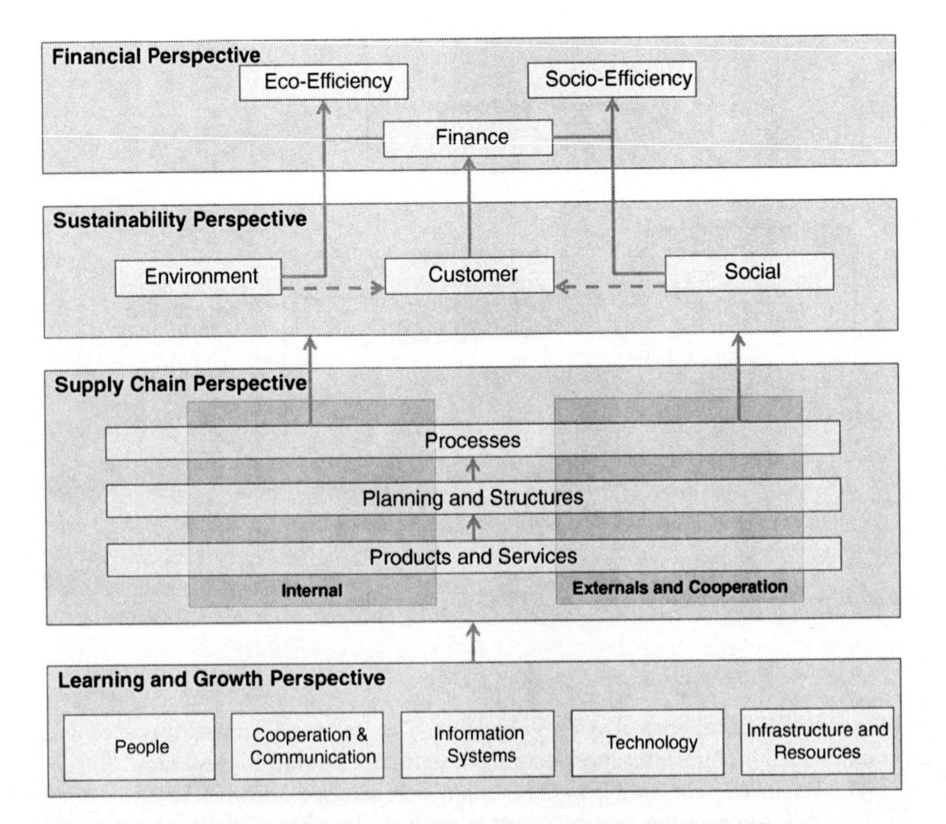

Fig. 2.14 Sustainable Supply Chain (SSC-)Scorecard – relations of the perspectives

2.4.6 SSC-Scorecard: The Financial Perspective

Actions into principle: Assess social and ecologic activities accurately from the economic point of view.

The SSC-Scorecard retains the financial perspective of the traditional version, since financial measures are valuable in summarizing the consequences of actions already undertaken (Fig. 2.15). Furthermore, sustainable financial results are the key objectives of all companies, and justify their existence in today's business environment.[20] This is also why this perspective forms the target framework for all other perspectives.

The SSC-Scorecard extends the traditional financial perspective with a further dimension related to environmental and social impacts: The (socio- and eco-) efficiency dimension. The idea behind this new dimension is to link the financial perspective with the other two sustainability dimensions. The aim is to assess social

[20]Except non-profit organisations.

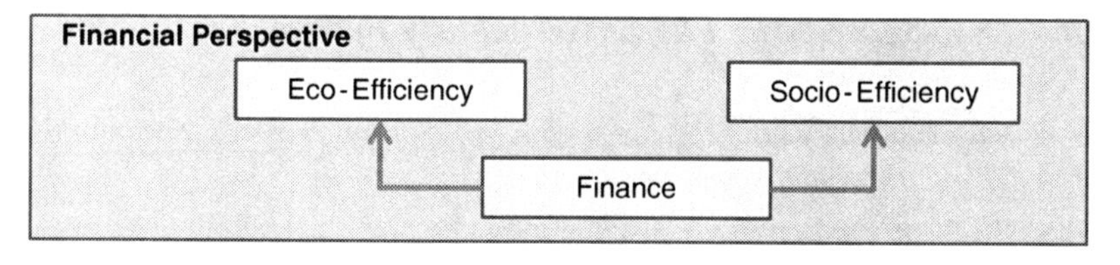

Fig. 2.15 The financial perspective of the SSC-Scorecard

and environmental activities from an economic point of view. Profitability and efficiency issues arise at the beginning and end of all activities involved in creating a sustainable supply chain, simply because activities must be financed and must provide a pay back after a specified time.

Bjorn Stigson, President of World Business Council for Sustainable Development (WBCSD), puts it this way: "This is what eco-efficiency is all about: combining the goals of business excellence and environmental excellence, and creating the link through which corporate behavior can support sustainable development".[21] This is one fundamental step to achieve the desired balance between economic, environmental, and social objectives, because compromises will only be offered or accepted by the people involved if the activity shows a high level of improvement while balancing environmental, economic, and social concerns.

The traditional measures for the financial perspective include Return on Capital Employed, Return on Investment, Return on Assets, Shareholder Value, Economic Value Added, and Market Value Added. Financial ratios influenced by supply chain management and logistics include Working Capital Efficiency, Operating Cost Reduction, and Fixed Capital Efficiency.

The basic calculation of the eco- and socio-efficiency measure is according to WBCSD:

= Product or Service Value / Environmental or Social Influence

Using this basic equation, companies can calculate eco-efficiency in a number of ways. The choice of indicators will depend on the needs of individual decision makers:

- A plant manager may wish to focus on the number of products shipped per kilojoule of energy consumed during manufacturing.
- A financial analyst may instead focus on the economic value of products sold per kilojoule.

The WBCSD therefore has developed a common framework for eco-efficiency indicators, with terminology consistent with the ISO 14000 series and the Global Reporting Initiative (GRI).[22]

[21] WBCSD (2005), p. 3.
[22] WBCSD (2005)

2.4.7 SSC-Scorecard: The Sustainability Perspective

Actions into principle: Link financial figures with customer needs. Extend this link by the environmental and social perspective.

The SSC-Scorecard retains a sustainability perspective, including the customer perspective of the traditional BSC (Fig. 2.16). "In the customer perspective of the Balanced Scorecard, managers identify the customer and market segments in which the business unit will compete and the measures of the business unit's performance in these targeted segments. This perspective typically includes several core or generic measures of the successful outcomes from a well-formulated and – implemented strategy. The core outcome measures include customer satisfaction, customer retention, new customer acquisition, customer profitability, and market and account share in targeted segments. But the customer perspective should also include specific measures of the value propositions that the company will deliver to customers in targeted market segments."[23]

The sustainability perspective of the SSC-Scorecard extends the customer perspective into the environmental and social performance dimensions. Both serve to measure the impact of your relevant supply chain on environmental and social performance KPIs. We have given performance measurement a separate chapter in this book, in which several such KPIs are described in detail.

One reason to put these two "new" elements into a single perspective together with the customer perspective is their potentially strong links. Customer and market segments of some industries, e.g. the retail sector, create direct environmental and social pressure. Hence, environmental and social responsibility performance may influence traditional core outcome measures such as customer satisfaction or new customer acquisition. Hence, the link serves to understand the impact of your sustainability performance on the competition from the customer perspective. Another reason to join these two elements within this "second"[24] perspective is to construct a causal relationship between the supply chain and the financial perspective. In both theory and practice it is desirable to measure the value proposition of a

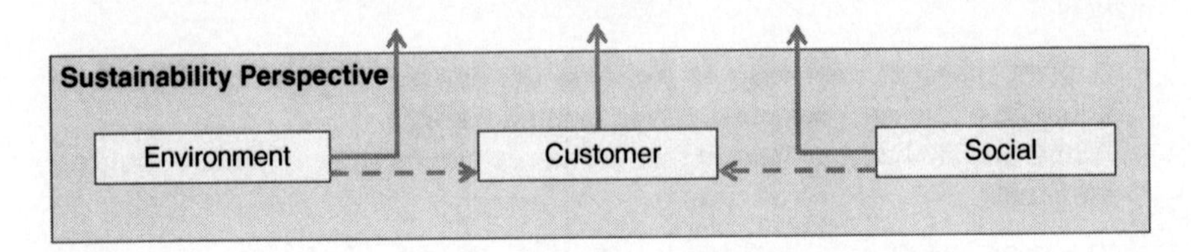

Fig. 2.16 The sustainability perspective of the SSC-scorecard

[23]Kaplan and Norton (1996).

[24]The perspectives are sorted in a sequence shown on Fig. 2.14, which represent the cause-effect relationships of the different perspectives. Hence, the sustainability perspective represents the second perspective.

supply chain at the company level, ideally in monetary terms. Some concepts do this very well, but only in financial terms, e.g. the Economic-value-added (EVA) model. The sustainability perspective between the financial and supply chain perspective extends these concepts by measuring the value proposition in environmental, social and economic terms, ideally linked to different supply chain resources (a “fourth perspective”).

2.4.8 SSC-Scorecard: The Supply Chain Perspective

Actions into principle: Incorporate tactical and strategic decisions along the end-to-end supply chain. Collaborate with your sales and product development departments and partners.

The SSC-Scorecard retains a supply chain perspective, including the internal-business-process perspective known from the traditional BSC. In the supply chain perspective, supply chain managers identify and improve the critical and key factors in which the supply chain must excel. Kaplan and Norton emphasize that only those process objectives should be established which serve to attract and retain the identified customers and which serve to achieve the defined financial goals.[25] The same requirement is valid for the SSC-Scorecard: Focus only on those supply chain factors that will have the greatest impact on customer, environmental and social goals, and consequently, on the defined financial goals.

Therefore the SSC-Scorecard extends the internal process viewpoint of Kaplan by another view, particularly relevant to supply chains: The External and Cooperation viewpoint. The justification for this is simple: External parties such as suppliers and logistics service providers affect (a) the internal processes of your company in terms of process capability and process reliability and (b) affect the sustainability performance of the whole supply chain; the supply chain you are part of or maybe, as an OEM, to which you respond from a stakeholder point of view. This differentiation in the SSC-Scorecard allows you to link the sustainability perspective and its associated sustainability performance indicators to internal and external *influence factors* in the supply chain perspective.

But what factors in a supply chain influence its sustainability performance? Let us look to the customer in the sustainability perspective. The customer judges the quality and service of a company based on the products and the service it gets; in logistics terms, for example, a late delivery. The environmental performance of a company is similarly the result of the physical operations of a company; in logistics terms, for example, the CO_2 emissions or the fuel consumption of a truck. Similarly with social performance, when we talk about traffic accidents, for example. Hence we can see that sustainability performance is directly influenced by operational processes in the supply chain perspective. This clear cause-effect relationship is

[25] Kaplan and Norton (1996), p. 26 ff.

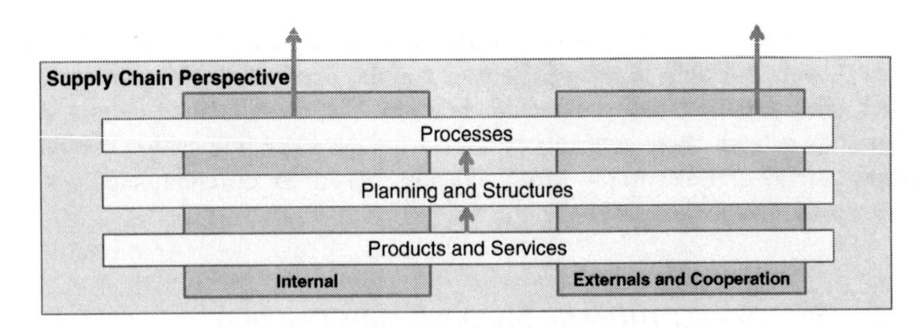

Fig. 2.17 The supply chain perspective of the SSC-Scorecard

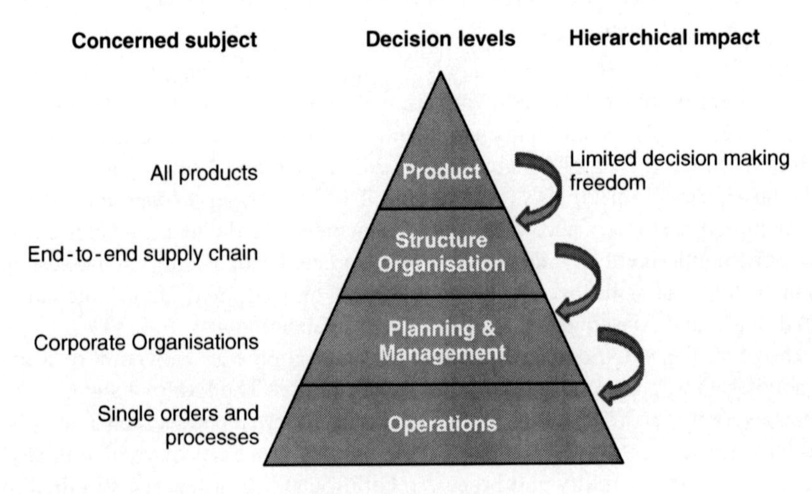

Fig. 2.18 Decision pyramid in supply chain management

reason enough to put the process dimension into the supply chain perspective within the SC-scorecard (see Fig. 2.17).

But should the supply chain perspective end at this point? Definitely not, because the impact of processes on the sustainability dimensions is mostly determined by previous supply chain planning and supply chain design decisions; and even by product design decisions (see Fig. 2.18). The shown four decision levels are different with regard to the time frame, scope, implications and deployment of capital and resources. Hence, the levels form a hierarchy and influence each other sequentially. Decisions at one level define the leeway for action with regard to further decision on the following levels, as decisions at a higher level curtail decision-making freedom at lower levels. Here is an example with regard to the environmental performance: A company decides to create a central warehouse strategy for its distribution operations. This structural decision constitutes a fixed parameter on the process level and has a key influence on the number of transport kilometres within the supply network; this in turn has a major effect on the volume of fuel required, or CO_2 emissions, by the company in question. Companies can obviously reduce overall fuel costs or CO_2 emissions by implementing optimisation policies at the process levels – by using route planning software, fleet control systems or alternative fuels, for example. But the decision with the broader scope and wider implications will have already been taken at the supply chain design level. Again, this clear cause–effect relationship is reason enough to put the "planning and structures" dimension into the supply chain perspective within the SC-scorecard (see Figs. 2.14 and 2.17).

Again the question: should the supply chain perspective end at this point? And again: No. According to the funnel-like logic shown in Fig. 2.18 all strategic and tactical decisions regarding planning and structures are, or rather should be, aligned to the products they supply, in particular with regard to product demand characteristics and, as already mentioned, with regard to product design. You will remember from Sect. 2.2 – on supply chain strategies – that different "demand" characteristics at product level impose distinctive supply chain planning and supply chain design requirements. Again, this clear causal relationship is reason enough to put the "product" dimension into the supply chain perspective within the SC-scorecard (see Figs. 2.14 and 2.17).

I will close this section with a final example showing the importance of these three interlinked dimensions within the supply chain perspective. The stipulated packaging dimensions of a product define the volume and the weight of the product and consequently the maximum number of products per load carrier – say per container load. This means that the decision on product design level influences the maximum theoretical utilisation of the space in a container, and therefore the capacity utilisation of a single transport vehicle, which would be assigned to the process level. The effects on the environment – in the form of CO_2 emissions per transported unit of the product – are therefore largely dependent on the decision on packaging dimensions at product design level. Actions on the planning and process levels such as route optimisation may naturally also have an impact on CO_2 emissions, but they have to be based on the fixed parameters of volume and weight and are therefore of lesser significance to environmental and resource protection.

2.4.9 SSC-Scorecard: The Learning and Growth Perspective

Actions into principle: Drive supply chain wide learning

The fourth and final perspective on the SSC-Scorecard aims to drive learning and growth in the supply chain. The SSC-Scorecard is in this respect firmly based on the traditional Balanced Scorecard of Kaplan and Norton. They postulate that objectives in this perspective enable the objectives and drive excellent outcomes in

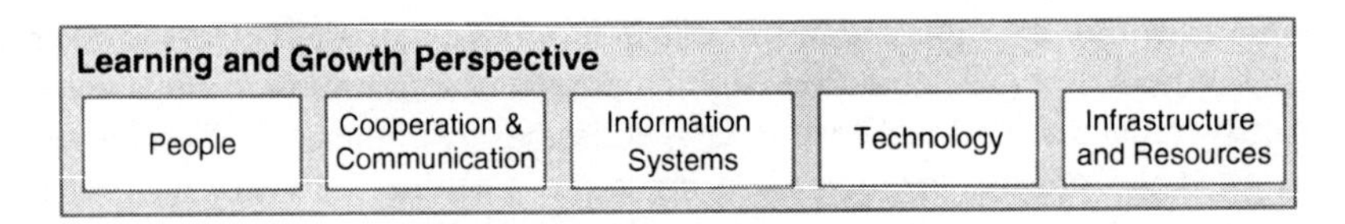

Fig. 2.19 The learning and growth perspective of the SSC-scorecard

the first three scorecard perspectives. Kaplan and Norton have defined three principal categories for the learning and growth perspective:

1. Employee capabilities
2. Information systems capabilities
3. Motivation, empowerment, and alignment[26]

These three categories are clearly also valid for sustainable supply chain development. However a sustainable supply chain strategy must extend the view along the whole supply chain and even beyond it. The learning and growth perspective of the SSC-Scorecards therefore comprises the following five principal categories (Fig. 2.19):

1. People capabilities
2. Cooperation and Communication: Trust, motivation, empowerment, and alignment
3. Information systems capabilities
4. Technology capabilities
5. Infrastructure and energy resources

The SSC-Scorecards extends the employee category of Kaplan and Norton, because the capabilities of people among your suppliers or service providers and even your customers deserve attention in a sustainable supply chain. Effective communication and cooperation based on trust and alignment of interests and objectives are essential to sustainable supply chains in the long term.

Furthermore, a sustainable supply chain strategy must pay attention to the development of technologies, infrastructure, and energy resources along the supply chain. Hence, the SSC-Scorecard extends the categories of Kaplan and Norton by the following:

(a) Technology capabilities: Technology plays an important role in today's supply chains and logistics. Transportation, warehousing, packaging, identification, and handling are increasingly impacted by technological developments. Interestingly many recent innovations in this area are about environmental protection, resource protection, social, and safety topics – all key ingredients of the sustainable development we are aiming for. Some technology examples

[26]Kaplan and Norton (1996), p. 127.

include: alternative fuels, hybrid powertrains, lane departure warning and automatic distance maintenance assist for trucks, renewable energy for warehouses, etc.

(b) Infrastructure and Energy Resources: Logistics and transport use infrastructure such as roads and energy resources such as fuel to operate. The SSC-Scorecard aims to monitor infrastructure- and energy resource-related data, which may affect the objectives of the three other SSC-Scorecard perspectives. For example, if a company aims to have reduced transport costs in its transport-intensive operations, the monitoring of fuel price and toll charges will deserve close attention. Examples of infrastructure and resource data include: infrastructure charging (e.g. tolls), speed and driver rest policies, oil prices, the price of emission certificates, infrastructure operator charges, the services and prices of service providers (e.g. a new intermodal connection from A to B), etc.

It is important in this scorecard-group not to create too many KPIs. The purpose of the infrastructure and energy resource grouping is to identify and then monitor those elements which may have a critical effect on the objectives of the following three perspectives. This Scorecard group is not intended support day-to-day operations as an information system.

The learning and growth perspective of the SSC-perspective feeds directly into the supply chain perspective, which is divided into an internal and an external perspective. Assignments split between the external and internal perspectives allow you to derive separate activities and to adjust your supply chain collaboration and business model according to the resources and capabilities available.

The learning and growth perspective is the last perspective of the SSC-Scorecard. The four perspectives and all their sub-groups represent a template. They are not carved in stone; rather, they are intended to serve as a generic model, to be adjusted to fit your own distinctive industry, company, and regional factors. To continue our six-step approach: Take your re-designed or new supply chain strategy, take this SSC-Scorecard, take the KPIs from Chap. 3, take the cases shown in this book onto the best practice platform, and start to put your new supply chain strategy into operation. Bear in mind that the SSC-Scorecard is more than a tactical and operational measurement system. It serves as a strategic management system, to help you manage the strategy over the long run.[27]

The next, final step is to implement the strategy and the SSC-Scorecard at a strategic level. You can focus on the ingredients and action items already mentioned in the previous section. The strategy development and integration approach ends with step 6. Nevertheless, the need for early adaptation of the supply chain to dynamic changes in the business environment requires an iterative application of this approach.

---

[27] According to Kaplan and Norton (1996), p. 10.

2.5 Principles into Practice

The following are some summary conclusions derived from the content of this chapter and from the bestLog research.

1. Be aware of goal conflicts in implementing, pre-define them, and communicate them to your team and your executive board.
2. Monitor your stakeholders and your supply chain input resources and the risks and chances associated with them.
3. Follow an holistic approach to define strategic gaps.
4. Drive supply chain-wide learning.
5. Link financial measures with customer needs. Extend this link by the environmental and social perspective.
6. Extend the responsibility of your supply chain managers (cross-company and cross-functional).
7. Consider sustainability as an opportunity and not only as a risk.
8. Communicate your plans, successes and sometimes, your failures internally and externally.
9. Get commitment from your top management.
10. Develop your own customised Sustainable Supply Chain (balanced) Scorecard.
11. Develop realistic targets for your SSC-Scorecard.
12. Clarify and translate the strategy and vision.
13. Enable feedback and learning circles in your organisation.
14. Establish an appropriate organisational framework and training resources.
15. Create an appropriate culture and awareness along your supply chain.

Tasks:

1) Can you write down, without checking company documents, the mission, strategy and supply chain strategy of your chosen organisation? Can you list any organisational strategic goals?
2) Is sustainability included into your chosen company strategy and/or supply chain strategy? Discuss how company actions at the strategic level influence sustainability (for example by: choice of supply chain design, selection of transportation mode, sourcing).
3) How does your chosen organisation communicate strategic goals to its employees and stakeholders?
4) Using example companies, analyse products shown in Table 2.1. What is their key characteristic: primary, functional, or innovative? How does this influence supply chain design?
5) Using Table 2.2 as an example, analyse the strategy of your chosen organisation at different levels. Are they interlinked? Discuss any contradicting goals and how you can manage such conflicts?

Bibliography

Aronsson, Hakan; Brodin, Maria Huge (2006): The environmental impact of changing logistics structures. In: The International Journal of Logistics Management, Vol. 17, No. 3, pp. 394–415

Brewer, Peter C.; Speh, Thomas W (2001): Adapting the balanced scorecard to supply chain management. In: Supply Chain Management Review, Vol. 5, No. 2, pp. 48–56

Bullinger H-J, Kühner M, van Hoof A (2002): Analysing supply chain performance using a balanced measurement method. In: International Journal of Production Research, Vol. 40, No. 15, pp. 3533–3543

Chopra S, Meindl P (2004): Supply chain management. Strategy, planning, and operation. 2nd edition. Upper Saddle River, NJ: Prentice Hall

Christopher M (2005): Logistics and supply chain management. Creating value-adding networks. 3rd edition. Horlow: Financial Times/Prentice Hall

Emmett S, Sood V (2010): Green supply chains. An action manifesto. Chichester: John Wiley

Esty DC, Winston AS (2006): Green to gold: how smart companies use environmental strategy to innovate, create value, and build competitive advantage. New Haven and London: Yale University Press

Fawcett SE, Ellram LM, Ogden JA (2009): Supply chain management – from vision to implementation. Upper Saddle River, NJ: Pearson

Fisher ML (1997): What is the right supply chain for your product? Harvard Business Review, 72: 105–116.

Kaplan RS, Norton DP (1996): The balanced scorecard: translating strategy into action. Boston, Mass: Harvard Business School Press

Kazem C, Richard L (2008): Sustainable competitive advantage: towards a dynamic resource-based strategy. East London Business School University of East London, UK

McKinnon, Alan C (ed) (2010): Green logistics. Improving the environmental sustainability of logistics. London: Kogan Page

O'Marah K, Hofman D (2010): The AMR Supply Chain Top 25 for 2010. Gartner Inc. doi: http://www.gartner.com/resources/201200/201212/the_amr_supply_chain_top_25_201212.pdf

Orsato, Renato J (2006): Competitive Environmental Strategies. When does it pay to be green In: California Management Review, Vol. 48, No. 2, pp. 127–143

Porter ME (1985): Competitive advantage. New York: Free Press

Porter ME (1987): From competitive advantage to corporate strategy. Harvard Business Review, 65: 43–59

Porter ME (2008): On competition. updated and expanded ed. Boston, Mass.: Harvard Business School Publ (The Harvard business review book series)

Sarkis, Joseph (Ed.) (2006): Greening the Supply Chain. London: Springer

Shrivastava P (1996): Greening business – profiting the corporation and the environment. Cincinnati, Ohio: Thomson

Straube F, Cetinkaya B (2008): Environment and logistics. Page 62–81, in Straube F, Pfohl H-C: Trends and strategies in logistics – global networks in an era of change. Bremen: DVV Media Group

Werbach A (2009): Strategy for sustainability: a business manifesto Adam Werbach. Boston, Massachusetts: Harvard Business Press

World Business Council for Sustainable Development (2005): Eco-efficiency. Learning module. Geneva: WBCSD

Chapter 3
Monitoring Performance

Wojciech Piotrowicz

Learning Goals.
By reading this chapter, you will:

- Understand the main metrics dimensions that reflect impact on sustainability
- Learn about the roles and importance of performance measurement
- Be aware of challenges related to creation of the performance measurement

3.1 Introduction

The previous chapter was about strategy and its implementation. Performance measurement and management are necessary to plan actions and monitor how a selected strategy is realised and the defined goals achieved. Performance measurement also provides information for decision-making at an operational level and reduces risks (both these topics will be discussed further in later chapters). This chapter is about the performance achieved as a result of the managerial decisions and strategies discussed earlier in this book. It is important to stress we will talk not only about the financial outcome generated by a company, but also about the wider results of all company activities, about all the outcomes and impacts that are associated with a company's existence and operations. Moreover, in this chapter we will concentrate on the performance of the whole supply chain from the point of origin to the point of consumption, beyond organisational boundaries and beyond traditional financial measurement. This chapter will deal with the evaluation of the supply chain according to its impact on sustainability: in economic, social and environmental dimensions, considering supply chain product, processes and outcomes. We will explain how performance measurement and evaluation can provide the necessary information to increase the positive effects of the supply chain while decreasing the negative impacts. However, the performance cannot be discussed if it is not evaluated and compared against defined objectives or the performance achieved by business partners or competitors (especially those named as best in the

C. Tyssen et al., *Sustainable Supply Chain Management*,
DOI 10.1007/978-3-642-12023-7_3,

industry). Performance also cannot be analysed without understanding the context in which the company operates, including the social and economic aspects. The most objective way of evaluation is measurement, which reflects performance in a quantitative way. However let's start from the question, why we really need to measure and monitor performance?

There are various reasons to use performance metrics, and here we need to stress the word "use". It is important to remember that measurement without defined purpose, just collecting data that will never be used, is only a cost for the company, a waste of human and financial company resources. Performance measurement is necessary to manage performance, to take actions based on data that reflects company performance and its external impact on sustainability. Collected information about past and current performance should be communicated to relevant parties and used. Company and overall Supply Chain performance should be analysed and improved, as Halachmi (2005) stated "Performance measurement is the only way of managing performance". The measures collected by the organisation should be reviewed on a regular basis by company staff, if necessary, shared with business partners and other stakeholders. This should help to improve overall supply chain performance, by improving performance of individual companies and processes, or reducing the negative effects on sustainability at all levels.

At the company level, performance measurement is developed and used in a more or less formal way. It is the everyday job of every single manger at every company to look at the processes and outputs under their responsibility. In a small company, there may be no need to use advanced technology and complex measurement systems, but in large organisations with complicated interlinked processes it is not easy to capture process performance. At the Supply Chain level, the situation is different, and more complex, as illustrated by the Supply Chain trends (Table 1.1). As supply chains include many independent organisations, individual metrics and sets of metrics are used by each organisation. Non-standard metrics definitions, different the period when data are collected or various data standards, create difficulties in measuring end-to-end Supply Chain performance in relation to economic, social and environmental dimensions. The typical performance measurement methods originating from traditional performance measurement approaches were designed for internal use and mainly to measure economic indicators, and were developed from managerial accounting and reporting. In most cases, sustainability measures are missing in such a measurement system. This is going to change. In Chap. 1, the negative external effects of Supply Chains were listed, such as the number of people killed on the road and Co_2 emission by sector or from the road transport (Fig 1.3–1.5). Such measures are linked to European and national efforts to reduce the negative impact of transportation-related processes. As a result, Supply Chain members need tools that can assist in impact reduction and the fulfilment of goals set by governments. The public and private sector need to find ways to assess current status, to plan and monitor improvement. As there is a shift towards a sustainability agenda, some of the traditional methods can be extended by environmental and social dimensions and adapted to a particular Supply Chain context. This is necessary as the lack of common approaches to measure supply chain impacts on sustainability creates problems for how to demonstrate to internal and external stakeholders that the company follows a principle of sustainable supply chain management, both at the company and at the supply chain Levels and across all chain members. The dimensions used to measure the impact on sustainability discussed here were developed during the BestLog project, through the analysis of existing best practices, existing literature and a number of workshops with practitioners. Furthermore, the measurement framework was tested and applied to look at the practices available in the Bestlog case studies (presented in this book and on the ELABestlog webpage).

3.2 Measuring and Monitoring Sustainable Supply Chains

Measurement that includes sustainability aspects is different than in a traditional, business oriented, performance measurement. When the sustainability dimensions are taken into account, the scope of the measurement should be extended. Sustainable development includes not only economic, but also social and environmental dimensions. Thus, the impact of the Supply Chain on such dimensions should be monitored. It is important to stress that ideally the impact of the WHOLE Supply Chain should be analysed, not just the partial impact of a single organisation. However, in most cases, it is impossible to measure the impact of such a complex and dynamic structure as the whole Supply Chain. Nevertheless there should be some attempt to look at the Supply Chain as one entity, composed of small blocks – individual companies and processes. This is important as in many cases, any negative impacts from a decision in one part of the supply chain could occur in a different place along the supply chain without the knowledge of the organisation where the original decision was made. While a company should look at its internal processes and performance, it is important to also look at how they influence the external world – business partners, the natural environment and society. The impact of supply chains on sustainability, and their relevant evaluation, takes place at several levels, from the bottom level such as individual actions and processes via operations to the strategic level, organisational-level performance and impact, up to the impact of the whole Supply Chain.

3.3 Evaluating the Impact of Your SC Activities on Sustainability

Within the BestLog project we developed framework that can help evaluate Supply Chain performance. This is built upon three main dimensions: social, economic and environmental, which are further sub-divided (Fig. 3.1). This modular composition allows performance to be analysed in a flexible manner, and such an approach can be even used in the small companies, when precise data may not be available.

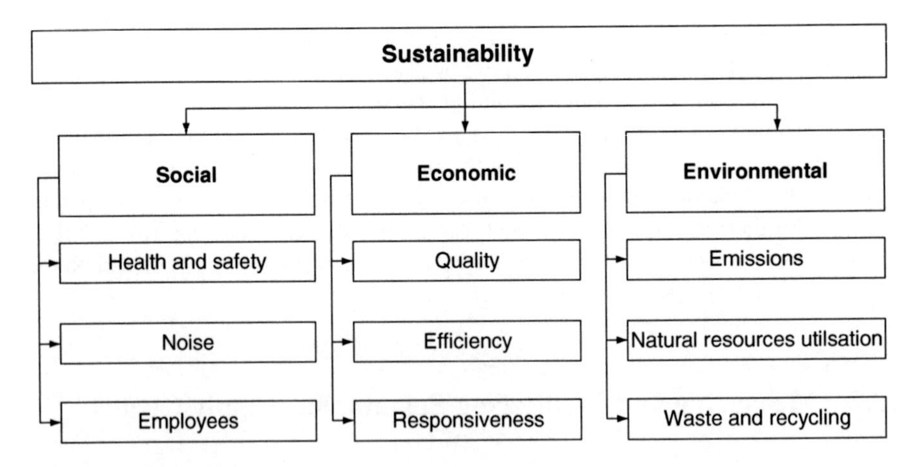

Fig. 3.1 Metrics dimensions and sub dimensions

3.3.1 Economic

Economic performance is a pillar of company activities. There is no doubt that a company should be profitable in order to be sustainable in the long-term. Practices which are named as "green" and "sustainable" but are not profitable in the long term cannot be considered as sustainable over time. We consider economic issues as the most important when looking at sustainability, as positive economic impacts increase the likelihood that certain process or activity can continue and be sustained. Within the economic impact we also include: quality, efficiency and responsiveness (Table 3.1) as they are necessary to be profitable in the long term. The quality of logistics services and customer satisfaction are key issues to generating long term profits. This can be measured by various metrics. To measure quality you can look at information available within the company (number of complains, delays, returns, damage, spillage, number of mistakes in documents etc.), as well as at data available from external sources, such as customer surveys to determine perceived service quality, feedback from business partners. However, as was discussed in the Chap 2, metrics should reflect company strategy and different customer segments. The same quality level may be acceptable for some customers, while not for a different group. Responsiveness, the second dimension which we defined, reflects how a company can respond to customer needs and a changing environment. In many ways it is related to the quality of service. Responsiveness can be measured by looking at product availability along the supply chain. When customer demand is changing, how fast can the company fulfil new customer needs, how fast is the process from product design to delivery, how long it will take to redesign distribution system, how accurate was demand forecasting? Responsiveness is to some extent in opposition to the last dimension – efficiency, as the creation of more responsive system requires underutilized capacity and not highly specialized process design. Thus, a balance between dimensions is needed.

Table 3.1 Economic dimension, sub-dimension and metrics examples

Economic dimension		
Sub-dimension (category)	Benefits and improvements achieved (examples)	Measures (examples)
Quality	Quality of products and services Customer service level Availability	No of stockouts, Customer response time, Product lateness, On-time delivery (%), Delivery reliability, Customer satisfaction, Customer complains, Product availability
Efficiency	Utilisation Productivity Cost reduction	Cash-to cash cycle, ROI, Inventory, level, total logistics cost, value added productivity, Asset returns, Inventory utilization, Truck fill rate, Warehouse utilisation. Delivery cost
Responsiveness	Response to customer needs Response to market changes Flexibility	Time to market, Production flexibility, Response time, Order cycle time, Track and trace performance, Order flexibility, Order fulfilment lead time

The efficiency dimension is the most familiar dimension for all managers. It is about cutting costs, increasing resource utilization, and process time reduction. This can be reflected by a variety of detailed measures, there are also attempts to use aggregated measures such the number of days supply chain cash-to-case cycle (Farris and Hutchison 2002).

3.3.2 *Environmental*

While the economic dimension is primarily about profits and the long term company existence, it is not enough to be sustainable in relation to overall company impacts. Thus the second dimension of impacts includes environmental issues. The environmental group of metrics focus on lowering the negative impact of transport and other company activities on the natural environment. The environmental dimension includes three sub-groups: emissions, natural resources utilization, waste and recycling that were defined during the BestLog fieldwork (Table 3.2).

Case:
Telematics at SMEs
Sieber introduced telematics system to reduce fuel consumption in their fleet by monitoring driving style
(See Case Collection in Part IV of this book)

Until recently, environmental aspects in typical measurement systems used by companies have been largely ignored. Thus, there is lack of commonly accepted

Table 3.2 Environmental dimension, sub-dimensions and metrics examples

Environmental dimension		
Sub-dimension (category)	Benefits and improvements achieved (examples)	Measures (examples)
Emissions	CO_2 emission Other pollutants emissions	CO_2 produced per litre delivered (grams), total CO_2 emissions in tons, reduction of CO_2 emission (% or in tons)
Natural resources utilisation	Fuel consumption Water consumption Land use Energy consumption	Reduction of fuel utilisation, reduction of water utilisation, land utilisation, energy utilisation per warehouse, reduction of need for road transport
Waste and recycling	Waste reduction % of materials/products recycled Bio-degradable materials used	% of packaging recyclable, % of packaging reused, reduction of equipment losses, reduction of cargo damage, reduction of spillages and leakages, reduction of obsolete, out of date items in warehouse

measurement norms and standards. Nevertheless when analysing company impact there is need to look how company activities influence the environment. This includes CO_2 emission, as there are EU and national targets to reduce CO_2. CO_2 measurement is at the development stage. There are different software packages and approaches and there is no universal agreement on which one is accurate and will be commonly accepted in the future. Energy use from the fossil fuels can be used as surrogate of CO_2emission. This also overlaps the sub-category of nature resources utilisation – fuel, water, land use. All such aspects are quite easy to measure. Data are available from the utility providers or fuel stations. Apart from CO_2, other categories of emissions should also be captured. Waste and recycling is another category related to the impact on environment. This can be reflected by the percentage of waste, and different categories of waste, sent to landfill in relation to waste which was recycled.

3.3.3 Social

Traditionally the social dimension is still not commonly incorporated directly into organisational performance measurement systems. Even if such measurement exists, it is often separated from performance measurement, for example incorporated into Human Resources department or section. However, people and their skills and their impact on society are key issues in sustainability. Social dimensions defined during the BestLog project include (Table 3.3):

- Health and safety
- Impact on employees
- Noise emission

Table 3.3 Social dimension, sub-dimensions and metrics examples

Social dimension		
Sub-dimension (category)	Benefits and improvements achieved (examples)	Measures (examples)
Health and safety	Toxic, hazardous emissions Accidents Working conditions	No. of employees/subcontractors/suppliers trained in health and safety procedures, no of accidents (within company), no of accidents that involve third parties, near misses, no of serious/fatal accidents, reduction of spillages and leakages (chemical substances), reduced accident probability
Employees	Employment Training Job security	% of production in EU, no of jobs created, no of workers trained to use new system, no of jobs reduced, no of people on long-term contracts, % of staff trained, no of drivers trained, staff retention, absenteeism
Noise	Volume Timing and location	Noise reduction (time of truck delivery), noise reduction from warehousing operation, activities switched/created at industrial zones

Health and safety incorporates metrics such as: The number and type of work-related accidents. In many countries there is a legal requirement to collect such data, especially in relation to fatal and major injuries, but again there is lack of standardisation and common definition across countries, even within EU. This category also needs to include all accidents which occur to sub-contracting and agency workers, not only internal employees. In addition all near-misses should be recorded to prevent accidents happening in future.

Employment can be measured at different levels, as overall job creation or reduction, and at the company level reflecting working conditions and employees morale. Such issues could be captured by measures like: level of absenteeism, number of new employees, percentage of key staff turnover (Gosselin 2005). A company also needs to consider the impact on business partners, especially small and medium enterprises. This can be reflected by metrics such as: % of transactions with SMEs, number of joint initiatives, and training provided for SMEs. As presented in the Chap 4, knowledge and training play important roles. However both are hard to capture. This could be reflected by the amount of training expenses (Gosselin 2005), % of staff with certain level of education or professional training and their changes over time.

Noise emission was also included into the social dimension, as it is less important in purely industrial areas. Noise emission is most disturbing in residential areas. Apart from the actual volume, the timing of any noise emission is also important, Time of the day or night.

3.4 Visible Supply Chain: Process, Product and Performance

To manage Supply Chain performance, looking solely at outcomes is not enough. Performance is influenced by two other interrelated elements: Process and Product (Fig. 3.2).

A visible, transparent supply chain allows you to consider the impact of every supply chain and every single member on sustainability. The impact is from both, products and processes. Product is the subject of the Supply Chain. Processes change the product, or support product delivery. Long term supply chain Performance measurement is influenced by both product and process, so in many cases trade-offs between both need to be addressed. However, this is not always possible, as the choice of products can be limited. Nevertheless there are options, especially for procurement managers, to select and source products by considering the product impact on supply chain sustainability. For example, local sourcing can reduce the need for transport. However, it is important to look not only at the distance transported, but also at the social dimensions. For developing countries, the export of raw materials or agriculture products can often be a major source of income. Although the distance transported can be long, trade with developing countries helps to stimulate their economic and social progress. Decisions about sourcing should be made carefully. The framework presented in the Fig. 3.1 can assist in making such decision.

Product impact on performance is related to sourcing, product flow monitoring, traceability and quality control. In certain industries, such as food processing, product is often traced from the field to the store.

However, even though there are interrelations between Processes and Products, from a measurement and monitoring point of view, they are treated to a large extent separately.

Only fully available information about products, processes and performance provides the ability to create a fully visible Supply Chain and capture the impact of the whole Supply Chain:

- Products – what is their origin, where they were treated?
- Processes – what was done to the products, how were the processes supported?
- Performance – what was the impact of both processes and products?

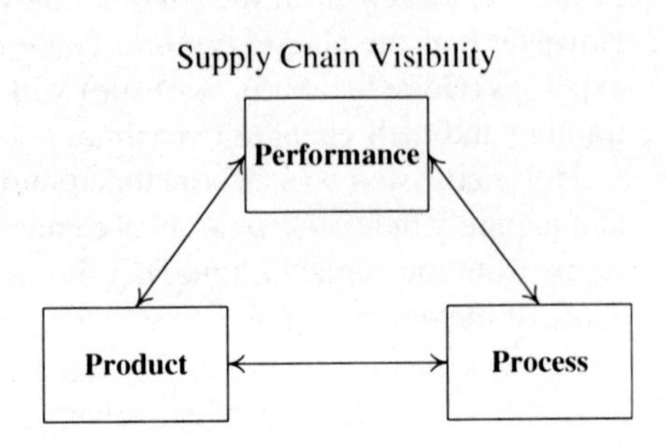

Fig. 3.2 Elements of visible supply chain

In the following sections, more about products, processes and performance will be presented.

3.5 Product Oriented Monitoring

There is a whole range of compulsory standards that concentrate mainly on products, their origin, flow and processing. They monitor, trace or audit product origin, processing and transportation. Such initiatives are required by legal frameworks, and developed at national and European levels. For example, they are typical for the food chain industry and give the opportunity to trace the path of products from the farmer to the shop. They were developed mainly as an answer to health-related problems and diseases among farm animals. Similar results are achieved by using product norms and standards. Voluntary codes of conduct and initiatives also exist. They could be for forestry products or fishing. They are mainly concentrating on assuring that products that enter the supply chain are from sustainable and legal sources. Even though product-oriented initiatives have various goals, they are built upon the same concept – to provide transparency of product origin and flow, assuring that products fulfil certain, and often strictly defined, criteria that are communicated and available to other supply chain members as well as to final customers.

There is also an issue related to products and their impact on the environment. While the fact that products are from sustainable and legal sources could be confirmed using auditing and monitoring techniques, there is the much more complex aspect of "embedded CO_2" – how much CO_2 and other gasses were emitted during product creation/growth. This varies and depend on complex factors, such as climate, location, manufacturing techniques, sources of energy.

3.6 Benefits and Use of Performance Measurement

Measurement should be more than just a simple exercise. It should generate benefits for the individual Supply Chain members and for the Supply Chain as a whole. Performance measurement can improve the understanding and collaboration between supply chain partners and increase Supply Chain integration. The use of common measurement metrics (which reflect performance of the process, or activity), methods and tools (which determine the approach to measurement data collection) allow consideration of the overall supply chain performance, and not just the performance of an individual member. Supply chain metrics could improve decision making processes, helping to define, test and implement new strategies (Chan et al. 2003; Gunasekaran et al. 2001). Performance measurement also supports other company improvement opportunities, such as targeting the most profitable market segments, service differentiation and cost reduction. At the Supply Chain level, performance measurement creates the opportunity to analyse the performance of the whole Supply Chain, as well as the impact of an individual organisation on the Supply Chain (Lambert and Pohlen 2001). The results of performance measurement can also be used to demonstrate the company impact on social and environmental dimensions, communicating to business partners and customers that the company has a positive impact or at least is trying to reduce the negative impacts from its operations. The following section shows how performance measures can be used and for what purposes.

3.6.1 Use Measures to Establish Current Impact and Performance

Performance measures can be used to establish the current impact of the company. While financial and other business-related data are likely to be known to managers, the social and environmental impacts may be unknown. When they are identified, then steps for improvement can be taken.

3.6.2 Integrate Long-, Mid- and Short Term Measures

Define measures for different time periods. Some initiatives to reduce the negative impact on sustainability need time and money. However, you can start from short-term goals and relevant measures. The implementation of short term measures can demonstrate progress and encourage further changes. It is better to define realistic goals and demonstrate that they can be achieved, then set unrealistic and too ambitious targets.

3.6.3 Use Performance Measures in Planning

Targets and goals should be reflected in numbers and values. Progress can be tracked over time and interim goals can be used to check whether the progress speed is as fast as was planned. Measures can also include scenarios, what-if questions, acknowledging that performance can depend on external factors.

3.6.4 Monitoring Strategy Implementation

Strategy implementation can be monitored using metrics (Key Performance Indicators – KPI) and trigger actions that help achieve defined goals, not only to measure economic progress, but also to monitor the impact society and the environment , as was already mentioned in Section 2.4.5 Performance metrics allow goals to be set, to track progress, and to initiate any corrective actions required to achieve the strategy defined. The monitoring of the current economic, social and political context may help to adjust the strategy, and relevant goals and metrics, to any new situation in which the company may operate.

3.6.5 Use Measures to Manage Processes

Process control and performance monitoring help to manage and improve processes. Measures related to processes, sub-processes and activities help to eliminate non-value added tasks and track improvement. It is important to notice that managers will often face a dilemma, as there are trade-offs between the economic, social and environmental dimensions, as well as between short- and long-term goals. However decisions should be made in line with the organisational strategy, or the strategy should be modified to reflect newly discovered opportunities revealed after process analysis.

3.6.6 Use Measures to Reduce Risks

In a later part of this book, there is a whole chapter related to risks. Performance measurement can be applied to reduce risks, both internal and external. Changes in performance, not fulfilling certain performance criteria can be used as an early warning to prevent problems or to avoiding the situation where small problems may become big ones. You can set the critical value of certain measures, which indicate increasing risks to your business. Link your measurement with your business partners. Early warning measures can help you to be better prepared for challenging times.

3.6.7 Motivation

Performance measurement can be used to motivate people, teams and business partners. As the goals, at individual, departmental and company level can be set and expressed in numbers, it is possible to monitor progress and motivate to achieve it. Of course when goals are completed it is time for reward and celebration.

3.6.8 Support for Training

Performance monitoring can indicate the weakest points in the system, which processes, functions or departments are struggling to achieve the defined performance. Detailed analysis of the causes can indicate that additional training is required. In addition, monitoring the training level of staff members from certain areas can assure that certain standards are met.

Case
Inditex decided to train 100% of employees to increase environmental awareness.
(See Case Collection in Part IV of this book)

3.6.9 Communicate to Employees and Enhance Responsibility

The company impact and the impact of individual processes can be communicated to employees by providing relevant measures. This can extend the responsibility of individual managers, as they will be able to see how their decisions influence not only economic, but also social and environmental, impacts.

3.6.10 Use Metrics for External Communication

Metrics and measures can be used as a communication vehicle. When company performance and impacts are known, when changes and improvement are monitored, results can be communicated to the external world, to stakeholders, NGO's and customers. When a company is accused of negative impacts, performance data that contradicts or explains such claims minimises any potential damage to the company reputation.

3.6.11 Segment Your Metrics

Performance measures should be specific. Different measures and performance values can be assigned to different processes, regions, customers and stakeholders groups. However, ideally there should be as large as possible overlap between measurement segments and areas. Measures should be as universal as possible. Even though values and goals may be different, they can be reflected by the same measure. This will allow to benchmark performance achieved in different parts of the organisation.

3.6.12 Know Where Your Products Come From

The importance of product tracing has already been mentioned. This is also relevant to the resources used by companies. Company reputation can be damaged when it is found that products are from suppliers which use child labour or are damaging the natural environment. Note that your different customers groups and markets can have a different "sensitivity" to particular issues.

3.6.13 Link Your Measures with Existing Standards

Different standards and certificates used in supply chain management (check www.elabestlog.org webpage for the full list of SC-related standards) include various measures that you can apply. The number of standards, or % of suppliers and business partners with certificates or standards, can also be used as a Measure. Remember to look beyond economic dimension, consider also social and environmental standards and certifications.

3.6.14 Standardise Metrics with Your Partners

Wherever possible try to standardise metrics with your business partners. Be sure that you are talking about the same things when you mention performance. Check how your business partners reflect performance, how they measure it and when. It will reduce problems if you standardise your metrics so that both organisations have the same view of performance in the same time period.

3.6.15 Benchmark Your Performance

Measures can be also used externally. The identification of key performance metrics is a major aim of any benchmarking process. When we know what performance level was achieved by the company (such as customer service level, fuel utilisation), it can be compared with the performance of organisations that are perceived as the best or best performing (Bhutta and Huq 1999). Benchmarking is most commonly used at the organisational or business unit level and is based on the assumption that it is important to learn from organisations that are perceived as the industry or sector leaders. Benchmarking could be operational, managerial and strategic (Shetty 1993). This allows comparison of the company performance at various levels and searching for ways to improve it. To select the best solution, managers can look at the outcomes of various initiatives, or their elements, and select the most promising ideas to be implemented in their organisations. Benchmarking can be also used to compare performance within the company or between various business units, to determine the best performers and learn from them – an internal benchmarking. However, the performance achieved by others should be reviewed carefully, as companies operate in different contexts. There are different social, political, economic, geographic factors that might influence the performance achieved. Issues such as company size, sector, structure and other organisational factors, influence results achieved, and should not be ignored. Looking at performance without context can result in misinterpretation, an unclear view between cause and effect, and the selection of practices that may not lead to improvement.

3.7 Problems with Measuring Performance

Measuring the whole supply chain is not an easy task, as in many cases it is unclear or unknown what should be measured, when and by whom. The international, often global, scope of Supply Chain operations, together with Supply Chain dynamics and a general lack of transparency make it difficult to define and visualise a Supply Chain structure, and it is hard to measure an undefined and unknown structure. Language barriers, as well as different data and IT standards, are other obstacles to measure end-to-end supply chain performance. Even though there are examples when the Supply Chain is measured and monitored from raw materials to end customer, such examples are limited to some companies, industries and specific situations. At the moment total end-to-end performance measurement across the whole supply chain is an exception, not a standard, and is mainly used in supply chains dominated by one strong organisation.

The second major problem of Supply Chain measurement is the lack of common measurement standards, approaches and tools. There is particularly a lack of accepted metrics related to environmental (where a good example is the question of how to measure CO_2 emission) and social issues across different markets. At the same time, companies have to deal with an overwhelming number of operational metrics used in various organisations used by different business partners. To create a successful supply chain measurement system, managers need to deal with various challenges at three levels: company, supply chain and context.

3.7.1 Problems to Measure Internal Performance

Even though performance measurement at the company level is more advanced than at the supply chain or contextual level, companies still have to cope with various challenges (Beamon 1999; Bullinger et al. 2002; Morgan 2004; Holmberg 2000). Metrics may not be related to strategy, may not provide a forward-looking perspective, may not be tied to effectiveness and efficiency, may be isolated and often changed, and may not be linked with customer needs. There is a domination of financial metrics that creates a misbalance between financial and non-financial perspectives (Gunasekaran et al. 2001). A further set of application problems can be identified, relating to the lack of links between strategy and performance measurement systems. This promotes a concentration on internal, not supply chain perspectives (Holmberg 2000). Moreover, there is a low level of differentiation between strategic, tactical and operational metrics (Gunasekaran et al. 2001), as well as between internal, organisational and supply chain metrics (Lambert and Pohlen 2001).

Additionally, managers have difficulty understanding metrics that are outside their functional area, and so naturally prefer to be evaluated based on outcomes that are under their control, while good supply chain management depends on inter-organisational cooperation (Brewer and Speh 2001). In many cases, small organisations do not have the time, resources and information required to analyse their activities and some of the medium sized organisations may have information, but do not have the skills to interpret this information (Morgan 2004). Large organisations, with highly computerised systems, are able to capture performance details of their operations, but internally only, or in relation to their closest business partners. However, in many cases such data are confidential and are not shared in the up or down-stream supply chains.

3.7.2 Problems at the Supply Chain level

As well as organisational factors at the company level, there are issues in performance measurement application that relate to the supply chain environment and its characteristics. In order to consider the performance of the supply chain, a company should deal with many complicated problems. Every organisation in the supply chain has a different corporate culture, policies and routines (Holmberg 2000), goals and objectives (Brewer and Speh 2001), while performance measurement systems generally ignore the different strategies of individual organisations (Beamon 1999). This creates challenges to overcome mistrust and suspicion between organisations in order to build and use the measurement system (Morgan 2004; Brewer and Speh 2001). Usage of internally oriented metrics, used incorrectly at supply chain level do not reflect supply chain performance (Lambert and Pohlen 2001) and metrics are not focused on process and cross-organisational perspectives (Bullinger et al. 2002). There is often a lack of standardised metrics used by all supply chain members (Brewer and Speh 2001), or that can capture performance of the whole supply chain (Lambert and Pohlen 2001).

Industry and sector differences are also influencing performance measurement in Supply Chain. Different sets of metrics are used for an industry or in a supply chain dominated by large company. There are industry-specific differences as certain metrics and performance standards are developed within the sector. For example, there are different characteristics for the automotive or food sectors. The supply chain model also influences the measurement and metrics selection. Different performance metrics will be important for build-to-order supply chains compared to traditional "push" supply chains.

Moreover, a single organisation could use various metrics for the same processes, as result of requirements from different customers or suppliers.

The complexity of supply chain activities impacts supply chain measurement (Lambert and Pohlen 2001) as well as the measurement being influenced by the supply chain model and its elements: stakeholder, collaboration level, business strategy, processes, and technology (Angerhofer and Angelides 2006). There are also related problems in systems implementation and usage, as information systems are often not able to capture supply chain performance at an inter-organisational level (Brewer and Speh 2001). These implementation issues are further complicated by the question of who will cover the initial costs of introducing a performance measurement system (Morgan 2004). Supply chain members use various IT systems, which often are not able to capture. Even though supply chain members might have internal performance measurement systems in place, there could be data duplication, incompatibility and integration problems between SC partners. As a result it may be difficult to create an integrated inter-organisational performance measurement system (Folan and Browne 2005). Finally, the lack of links between the rewards and recognition of managers and their supply chain performance (Brewer and Speh 2001) is another difficult application problem identified in the literature.

3.7.3 Problems to Capture Context and Its Dynamics

Performance is related to context and time. Context is dynamic and influences the performance achieved by organisations. To fully understand organisational performance, Social, Technological, Economic, Environmental and Political factors should be captured (and ideally quantified) as well as their changes over time. This helps both the selection and usage of performance systems, as well as providing a greater understanding of the performance achieved, to establish links between context and performance, between drivers, actions and outcomes. Looking at performance without capturing the context and moreover, the changes in the context, is pointless. Without understanding the context, it is impossible to distinguish between cause and effect and to understand factors that influenced the decision making process. Only when the context is defined, it is possible to look at generic learning and consider the possibility to transfer certain practices or solutions across industries, countries or sectors.

3.8 Stages of Supply Chain Performance Measurement

So the question is where to start building the performance measurement system? It takes time to develop a system that will measure company performance. In fact, companies are at various stages of performance measurement implementation (Fig. 3.3). Analysing case studies we found that some companies measure only their core processes, other are sharing metrics with business partners, while some large organisations are able to look at their end-to-end supply chain. However, there is a tendency for performance measurement systems to evaluate from intra- to inter-organisational scope (Folan and Browne 2005). Traditionally, performance measurement was concentrated in internal company activities, often limited to core operations. Some companies will not go beyond this point. In the case of many SMEs, achieving this level can be enough as their processes are easily visible and their performance can be easily captured without advanced performance systems. Other groups of companies monitor all their processes, both core and supporting. An extended performance measurement system looks beyond company borders, but often includes only the nearest supply chain partners, such as suppliers and

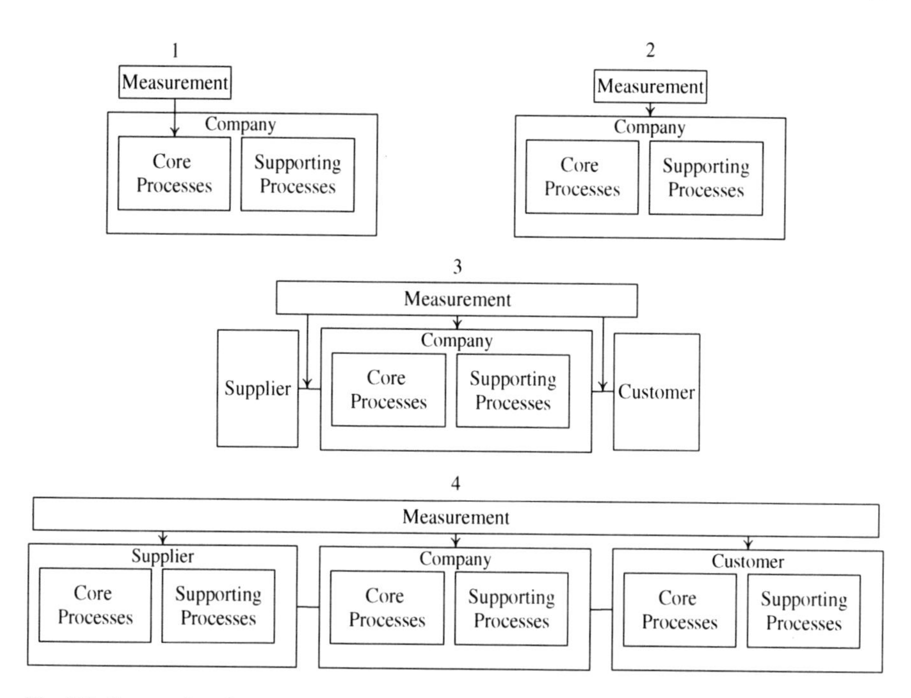

Fig. 3.3 Stages of performance measurement: (1) core processes only. (2) all company processes. (3) company processes and links with suppliers and customers. (4) SC performance measurement (includes all SC partners)

customers, and often only on the point of contact, or on the shared processes that link two companies – Supplier performance, or customer satisfaction are examples of such measures. Supply Chain Performance Measurement (SCPM) covers all the flow from raw materials to the final customer. Creation of the SCPM system (SCPMS) can be achieved by an evolutionary approach, linking systems used by individual companies, starting from standardisation and the exchange of selected key metrics, or in a revolutionary centralised way (creating the system at the same time), which requires a dominant company or industry association that can force, or finance, the creation of the supply chain system, which covers measurement across different organisations.

Case:
Sharp Electronics
Smart Way is a voluntary organization lead by Sharp Electronics, which aims on reducing Co2 emission and air pollution.

(See Case Collection in Part IV of this book)

In practice, again there are some limitations to measuring the whole Supply Chain and certain elements are excluded (for example customer is defined as a shop, or dealer, not an individual who buys the product). However, such simplifications might be necessary from a practical point of view. For the company, it could be a natural evolution from concentrating on core measures, through internal performance measurement, via a solution that covers part of the supply chain, to measuring the whole end-to-end Supply Chain. Looking at the supply chain performance from the perspective of a single company only, or relationships between two members could result in unintended consequences, when improvement at a single Supply Chain point could result in negative impact for the whole supply chain. This strategy is used by some companies that "move" some of their costs into suppliers, similarly use of a large number of subcontracting parties in countries without strict legal controls may hides some practices that could be disputed from the social responsibility or environmental point of view. In some cases, there could be companies that might not willingly participate in initiatives that increase Supply Chain visibility, as they already know that their, or their business partners, impact on sustainability is negative.

Case:
Siemens e-Integration
Siemens introduced a compulsory supplier evaluation system which looks also at environmental and social aspects.
(See Case Collection in Part IV of this book)

3.9 Supply Chain Performance Measurement: Methods and Approaches

A wide range of methods have been developed, but only some of them are accepted by practitioners and implemented to measure Supply Chain performance, among them the SCOR model and the Balanced Scorecard are common. The SCOR model, which is presented in the Chapt 4, includes a set of predefined metrics that can be used by a company and is focused on the company point of view. The second well known approach is the Balanced Scorecard (BSC) (Kaplan and Norton 1992, 1993, 1996). BSC looks beyond traditional financial performance measures. In the Balanced Scorecard metrics are grouped into four perspectives: customer, financial, business processes, growth and learning. In practice, BSC is a flexible tool, so it is often modified, and its perspectives are renamed or new perspectives are added (such as environmental dimension), to reflect a situation specific to the organisation. The Balanced Scorecard can also be modified to incorporate sustainability dimensions, which was demonstrated in the Chap. 2 where the Sustainable Supply Chain Scorecard was presented.

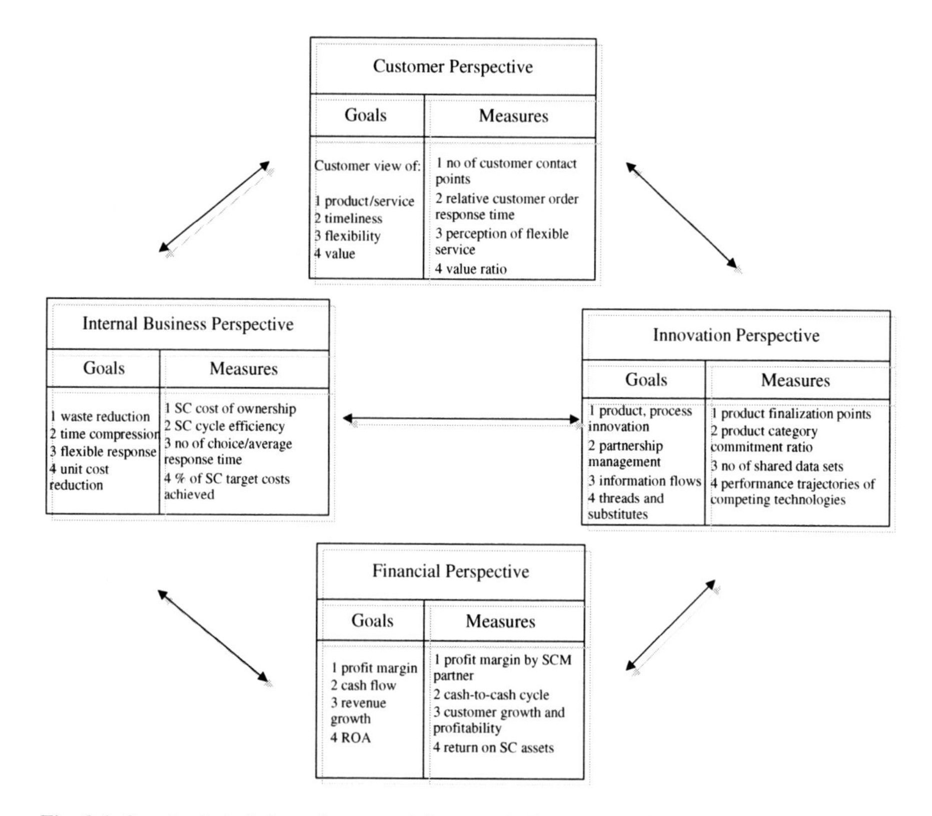

Fig. 3.4 Supply chain balanced scorecard framework (Brewer and Speh 2000)

3.10 Creating Supply Chain Performance Measurement System

As we stated before, there are both organisational and supply-chain related challenges to creating measurement systems, especially a system that will be able to capture the impact of the individual company and the overall supply chain. This section will overview what steps can be taken to create a supply chain performance measurement system. The best way is to start in your own organisation, to determine what metrics are already in use and by whom. It is possible to do this by adopting an approach based on process based measurement, which can be created by decomposing a process into sub-processes and activities and assigning to each of them metrics related to cost, time and outcome, Fig. 3.5 (Chan and Qi 2003).

Process decomposition requires stages such as:

1. Identification and linkage of all inter- and intra-organisational processes
2. Definition of core processes
3. Derivation of missions, responsibilities and functions of core processes
4. Sub-processes decomposition

Table 3.4 SC metrics classification based on balanced scorecard figure (Bullinger et al. 2002)

	Financial perspective	Customer perspective	Organisational perspective	Innovation perspective
Supply chain perspective	Total supply chain cost, total SC inventories, total SC revenues	Point of consumption availability, point of consumption product quality	SC relationship quality, productivity loss, perfect order fulfilment	Market share, new product time-to-market, new product time-to-first make
Process perspective	ROI, return on capital employed, cash-to-cash cycle, revenues	Customer satisfaction, customer loyalty, customer complains	Forecast accuracy, planning process cycle, schedule changes	% sales from new product, percent employees in cross-functional teams
Function perspective	Material acquisition cost, inventory cost, work in progress, cost per unit produced, freight cost, picking cost, transportation cost, cash flow	On-time delivery, order fill rate, order cycle time, invoice accuracy, number of bank orders, % resolution on first customer call, order trace and track performance	Incoming material quality, inventory count accuracy, out of stocks, line item fill, inventory turns, EDI transactions	Number of employee suggestions

5. Derivation of responsibilities and functions of sub-processes
6. Identification and decomposition of main activities
7. Create links between processes, activities and their goals

This exercise can then be repeated to identify the performance measures required by customers, suppliers and stakeholders for the inter-organisational processes, including communication with groups of customers and stakeholders.

Lohman et al. (2004) provided some advice as to how to design such a system, building it from a pilot stage, including:

1. Create a metrics dictionary, which includes metrics from operations, warehousing, transportation and customer service
2. Define standard metrics template to produce values in consistent way
3. Cluster metrics where it is possible
4. Use real data during the tests, to get feedback from users
5. Be sure that staff have allocated time to work on system development
6. Select software to build the system at the final stage of design

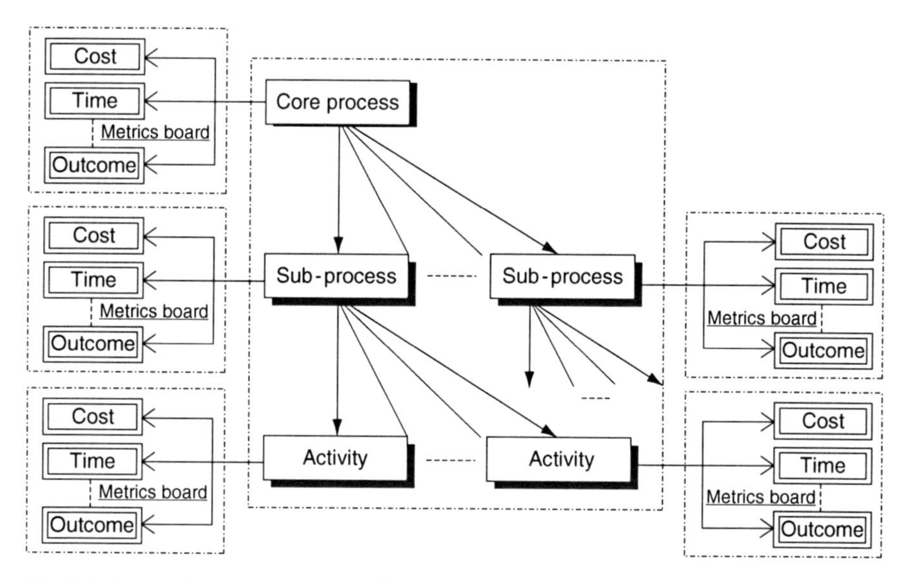

Fig. 3.5 Process-based measurement (Chan and Qi 2003)

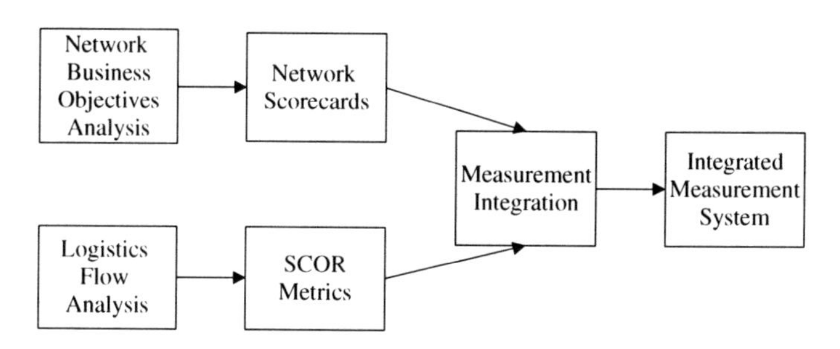

Fig. 3.6 Linking BSC and SCOR metrics (Bullinger et al. 2002)

This can then be applied to the whole supply chain. However, supply chain partners should cooperate from the earliest stage to agree performance metrics that will be used jointly and shared across the chain. Together with information about products this will create the opportunity to build a visible supply chain and demonstrate its impact on sustainability.

There is also the possibility to link the balanced scorecard and the metrics included in SCOR model (Fig. 3.6) (Bullinger et al. 2002). A supply network scorecard was linked and integrated with SCOR metrics, which control the logistics performance. The creation of such a performance measurement system requires supply chain analysis that includes the identification of measurement goals and the activities associated with them.

To achieve the goals (Bullinger et al. 2002), it is proposed to complete activities such as:

- Identify business drivers, operational indicators and existing measurement instruments
- Define balanced measurement based on business strategy and derived objectives
- Measure and document present performance levels
- Benchmark versus theoretical limits or others in/out the own organisation
- Evaluate sensitivity of changes in business drivers and operational indicators

Analysis of measurement is completed after identifying the main business objectives and the creation of supply chain transparency. The last part of this analysis is a conclusion, where improvement opportunities are identified and metrics developed. In addition, it is important to remember that metrics are not static. They can be changed naturally, as they are used, or in more radical way, for example when the company strategy is changed – as metrics need to reflect new strategy (Bourne et al. 2000).

3.11 Conclusions

This chapter presented the need for performance measurement, as measurement is a necessary activity to manage and improve performance. In today's world, managing solely financial performance is not enough – wider social and environmental impacts should be considered. Even though there are challenges in creating a performance measurement system that will cover all supply chain members, companies should start to create it by starting from their own internal performance and extending it to their closest business partners. Changes in tools and techniques allow the incorporation of sustainability dimensions. A performance measurement system creates a valuable source of information to manage internal and external issues, and the following chapters present such management activities. Monitoring performance is different from reporting and auditing. Information that is not used for improvement is just waste of your time and company resources. So use the metrics to manage company and improve performance, improving your profit and reduce negative impacts.

Tasks:

1) Using Metrics dimensions and sub dimensions Fig. 3.1 and metrics examples tab. 3.1–3.3, can you identify the impact of your area of responsibility on economic, social and environmental dimensions?
2) Can you repeat task above for the whole company? How do your actions and decisions influence the impact and performance of the whole organisation? Think about the impacts within company, impacts on business partners and customers, and finally on external impacts.
3) In the previous tasks, you identified the impacts of areas under your responsibility and the impacts of the whole company. Can you identify the impact of the whole Supply Chain in which you participate?

4) Review the case studies at the end of this book (you can use also cases from www.elabestlog.org). What categories of benefits and performance measures are reported there? Are similar measures used by your organisation?
5) Can you list the key measures used in your organisation? How important are they for internal use, for customers and business partners? Do they reflect all sustainability dimensions (economic, social, environmental)?
6) How are metrics collected by your organisation used? Do you need all of them? Can you simplify your performance measurement system?
7) Do you know the origin of products which enter your chosen supply chain? How important is product traceability in this supply chain? Discuss.

Bibliography

Angerhofer, B.J., Angelides, M.C. (2006): A model and a performance measurement system for collaborative supply chains. Decision Support Systems. 42(1): p. 283–301.

Beamon, B.M. (1999): Measuring supply chain performance. International Journal of Operations & Production Management. 19(3–4): p. 275–292.

Behn, R.D. (2003): Why measure performance? Different purposes require different measures. Public Administration Review. 63(5): p. 586–606.

Bhutta, K.S., Huq, F. (1999): Benchmarking – best practices: an integrated approach. Benchmarking: An International Journal. 6(3): p. 254–268.

Bourne, M., et al. (2000): Designing, implementing and updating performance measurement system. International Journal of Operations & Production Management. 20(7): p. 754–771.

Brewer, P.C., Speh, T.W. (2000): Using the balanced scorecard to measure supply chain performance. Journal of Business Logistics. 21(1): p. 75–93.

Brewer, P.C., Speh, T.W. (2001): Adapting the balanced scorecard to supply chain management. Supply Chain Management Review. 5(2): p. 48–56.

Bullinger, H.-J., Kuhner, M., van Hoff, A. (2002): Analysing supply chain performance using a balanced measurement method. International Journal of Production Research. 40(15): p. 3533–3543.

Chan, F.T.S., Qi, H.J. (2003): Feasibility of performance measurement systems for supply chain: a process-based approach and measures. Integrated Manufacturing Systems. 14(3): p. 179–190.

Chan, F.T.S., et al. (2003): A conceptual model of performance measurement for supply chains. Management Decision. 41(7): p. 635–642.

Cuthbertson R, Piotrowicz W (2008): Supply chain best practices – identification and categorisation of measures and benefits. International Journal of Productivity and Performance Management. 57(5): p. 389–404.

Cuthbertson R, Piotrowicz W (2009): Transferring best practices, in: Best Practices in Logistics and Supply Management. K, Rutkowski (ed.). Warsaw School of Economics-Publishing, Warsaw.

Farris II, M.T., Hutchison P.D. (2002): Cash-to-cash: the new supply chain management metric. International Journal of Physical Distribution & Logistics Management. 32(4): p. 288–296.

Folan, P., Browne J. (2005): A review of performance measurement: Towards performance management. Computers in Industry. 56: p. 663–680.

Gosselin, M. (2005): An empirical study of performance measurement in manufacturing firms. International Journal of Operations and Production Management. 54(5/6): p. 419–437.

Gunasekaran, A., Patel, C., Tirtiroglu, E. (2001): Performance measures and metrics in a supply chain environment. International Journal of Operations & Production Management. 21(1–2): p. 71–87.

Halachmi, A. (2005): Performance measurement is the only one way of managing performance. International Journal of Productivity and Performance Management. 54(7): p. 502–516.

Holmberg, S. (2000): A system perspective on supply chain measurement. International Journal of Physical Distribution & Logistics Management. 30(10): p. 847–868.

Kaplan, R.S., Norton D.P. (1992): The balanced scorecard – measures that drive performance. Harvard Business Review. 70(1): p. 71–79.

Kaplan, R.S., Norton D.P. (1993): Putting the balanced scorecard to work. Harvard Business Review. 71(5): p. 134–147.

Kaplan, R.S., Norton D.P. (1996): The balanced scorecard: translating strategy into action. First ed., Boston, MA: Harvard Business School Press.

Lambert, D., Pohlen, T. (2001): Supply chain metrics. International Journal of Logistics Management. 12(1): p. 1–19.

Lohman, C., Fortuin, L., Wouters, M. (2004): Designing a performance measurement system: a case study. European Journal of Operational Research. 156(2): p. 267–286.

Morgan, C. (2004): Structure, speed and salience: performance measurement in the supply chain. Business Process Management Journal. 10(5): p. 522–536.

Shetty, Y.K. (1993): Aiming high: competitive benchmarking for superior performance. Long Range Planning. 26(1): p. 39–44.

Chapter 4
Managing Within Your Organisation

Thorsten Klaas-Wissing and Christoph Tyssen

Learning Goals.
By reading this chapter you will:

- Learn about the general connection between the management of internal affairs, sustainability and logistics best practice
- Get to know the relevant management fields reflecting the internal enablers of sustainable Supply Chain Management (SCM)
- Be able to understand and apply the way how to select or prioritise the internal enablers of sustainable SCM
- Be capable of analysing and critically reflecting your own situation compared to logistics best practice

4.1 Under Your Managerial Control

After dealing with issues of strategy and performance in the context of sustainable SCM in general, in the preceding two chapters, we will now shed some light on the best practice of those relevant management fields of sustainable SCM, which are typically prone to be more or less under direct influence of the companies involved in a supply chain. Since the usual sphere of direct managerial influence lies within the boundaries of a company, we intentionally labelled this chapter "under your control – managing inside your organisation". Intentionally we arrange this chapter before "managing outside your organisation" (see next Chap. 5), because many experienced practitioners told us that especially best practice companies have learned to develop managerial excellence regarding their own operations first before going beyond the companies boundaries, trying to actively manage the supply chain of related suppliers and customers – simply expressed by the well known motto: *Get your own house in order before you go tearing others' houses down.* From our countless discussions with hands-on practitioners, general industry experts, association representatives, academics, and respective politicians, which we conducted in the dedicated workshops, via the bestLog online platform and in

C. Tyssen et al., *Sustainable Supply Chain Management*,
DOI 10.1007/978-3-642-12023-7_4, © Springer-Verlag Berlin Heidelberg 2011

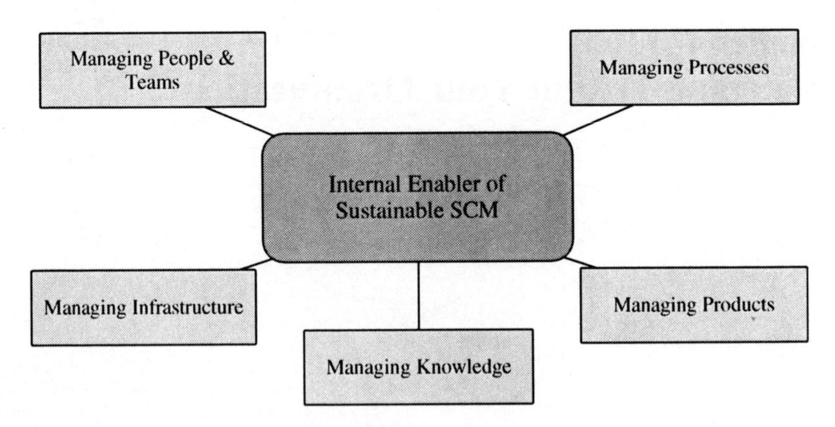

Fig. 4.1 Internal Enabler of Sustainable SCM

the case collection interviews during the bestLog project, we have identified 5 distinctive internal management fields, which have figured out to converge to major internal levers or enablers of sustainable SCM. The following Fig. 4.1 gives an overview of these respective management fields.

Though obviously being rather general in nature, the identified management fields offer direct levers for the establishment of sustainable SCM within and between companies. Initially, we were surprised about these results, but we had to learn especially from our manifold talks with experienced practitioners and appropriate academics, that sustainability is not a management task sui generis, but has rather to be interpreted as a cross-functional/cross-company philosophy or a common supply chain spirit, which should pervade all relevant fields of general management and SCM, respectively. By this, the responsible managers are essentially enabled to integrate concerns of sustainability into their day to day business decisions, as they will manage (i.e. design, plan, organise, implement, supervise and control) *processes*, *products*, *knowledge*, *infrastructure*, and *people & teams* accounting for the overall balance of social, economical, and environmental objectives. However, as a direct consequence of their general and pervasive character, the management fields of sustainable SCM challenge managers to select which of the internal enablers should be in focus, respectively. Since we cannot manage all these things at once there has to be an applicable guideline which helps to identify, evaluate, prioritise and finally select the respective enablers to be set in place for each company specific situation (Fig. 4.2).

The next five Sects. 4.2–4.6 will consecutively deal with each of the aforementioned management fields, following a rather straightforward "easy to read" composition. In line with the major goal of bestLog, the "promotion and dissemination" of best logistics practice, we thereby avoid and deliberately abstain from scientific elaborateness in order to assure comprehensibility of a preferably broad and practically oriented target audience. Hence, each section will give a short description of the particular management field with regard to its basic domain(s), management tasks, related management concepts and their respective relations to

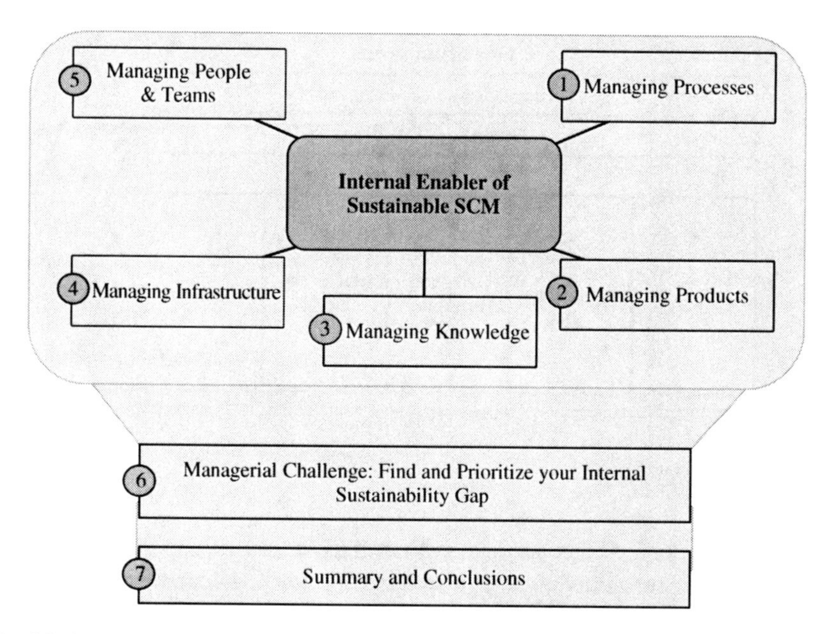

Fig. 4.2 Structure of the chapter

sustainability. Subsequently specific guidelines of enforcing sustainable SCM are presented, which can be understood as the quintessence or concentrate of the bestLog learnings, reflecting the experiences of those companies showing a best logistics practice. These findings will to some extend be illustrated by sensible extracts of the bestLog case collection in dedicated text boxes. In Sect. 4.7 we will finally present a simple check list that could appropriately help to analyse your current situation, find out respective gaps and prioritise the relevant management fields in order to achieve a more sustainable SCM. Section 4.8 will close this section by summarising the main findings and giving some overall conclusions.

4.2 Managing Processes

In general, every company consists of numerous (sequentially) interlinked business processes, such as procurement, production, distribution or logistics processes, which directly or indirectly contribute to its overall (economical, ecological and social) performance. Thereby, processes represent multiple series of logically and timely interlinked operative and administrative activities adding value to a (intermediary or final) product or service. The strategic importance of processes as enabler for sustainable competitive advantage has been stressed by the well known value chain approach of Michael E. Porter which gives some interesting hints in order to structure internal processes (Fig. 4.3).

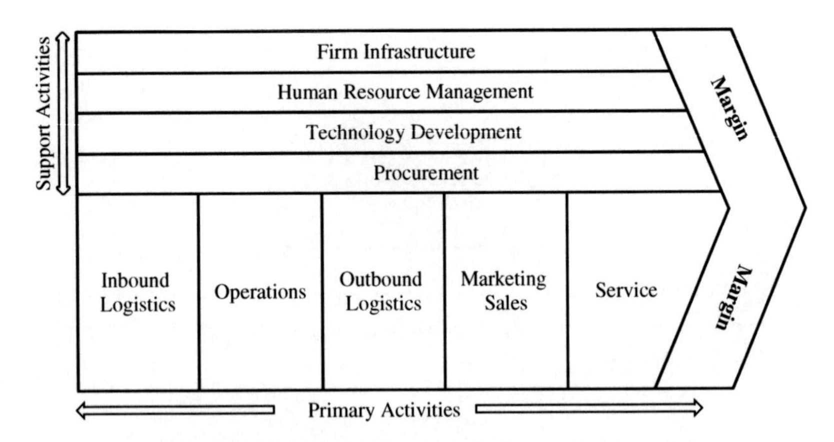

Fig. 4.3 Value chain by Porter (1985)

This value chain differs between primary activities and support processes and activities. Activities like human resource management, procurement and technological development rank among the support activities, whereas production, procurement and distribution represent primary activities. Each activity can be a strong enabler in terms of competitive advantage and can therefore – in our terms – also be strong enablers in terms of sustainable SCM within a company. Additionally, being part of a supply chain, companies are connected by inter-company processes, which eventually affect the performance of overall supply chains. Here also Porter offers the extended value chain of inter-company networks (Fig. 4.4).

In general, against the background of intensified competition, higher customer demands, rising cost pressure, and increasing environmental and social awareness of developed societies, managing processes in a well considered and systematic way is generally considered a "must have" competence of best practice companies. Hence, it is not surprising for us that all of the companies involved in the bestLog inquiry for best practice cases somehow deal with the management of processes, albeit in more or less sophisticated ways.

Managing processes encompasses the tasks of intentionally planning, designing, organising, controlling and performance-monitoring of business processes. Therefore, the term process management denotes the application of specific methods, tools, techniques and systems that visualize, define, measure, control, report and improve processes in order to profitably meet customer requirements. Thereby, process management aims at increasing information transparency, reducing physical and informational interfaces and establishing flow consistency in order to improve the overall efficiency of a value network. The typical outcome of process management is faster processes inducing lower costs and less use of resources. However, originally the concept of process management focused on mere economic goals. Nowadays, additionally considering ecological (e.g. carbon emissions induced by a process) and social (e.g. working conditions induced by a process) aspects within Process Management will lead to a sustainable company

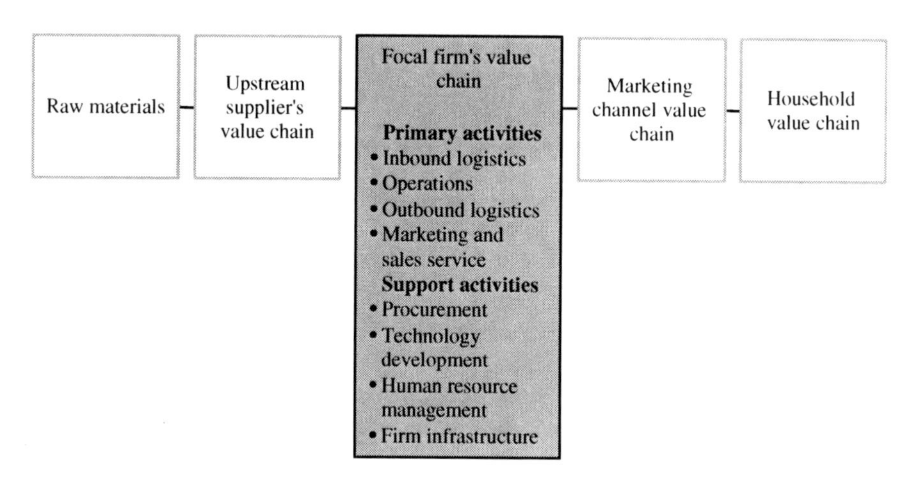

Fig. 4.4 The value system (Porter 1985)

performance and sustainable supply chain performance, respectively, in the best-Log sense of best logistics practice.

This means that process management will be successful especially by taking a *holistic view* of the company and the respective supply chain context(s) which the company is involved in. Furthermore, the application of a *total impact approach* to process management, i.e. considering economic (e.g. service levels and costs), environmental (e.g. emissions) and social (e.g. working conditions) aspects, bears significant potential with regard to sustainable process design solutions. Eventually, integrating *customer oriented thinking* – in the sense of internal as well as external customers – is pivotal to the establishment of consistent and transparent flows of materials, products and related information. As a consequence, managing processes in a well considered and systematic way is an eminently strong enabler for the establishment, maintenance, and improvement of sustainable business concepts in logistics and supply chain management. Against this background and in view of the multiple expert discussions, as well as the besLog case collection the following six guidelines for "managing processes" can be derived, which basically reflect the experiences of those companies showing a best logistics practice.

4.2.1 Know Your Processes

The first step on the way to sustainable process excellence is to be familiar with what your company is currently doing in your supply chain. Best practice companies therefore know their processes as they map and continuously update at least their critical supply chain processes, as well as the relevant process interfaces to customers and suppliers by using meaningful reference models. Next to

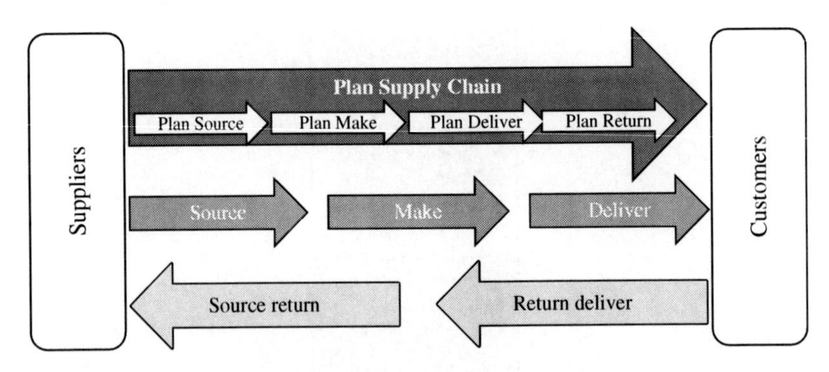

Fig. 4.5 Structure of the SCOR-Model (the supply chain council at www.supply-chain.org)

others, like the previously presented value chain approach, one prominent example of such a reference model is the Supply Chain Operations Reference (SCOR) Model (see Fig. 4.5) developed and regularly updated by the Supply Chain Council (see: www.supply-chain.org). This model provides a standardised methodical toolset, based on the simple process logic of *plan*, *source*, *make*, *deliver* and *return*, in order to map all relevant supply chain processes and the interrelationships of potential actors in a supply chain, such as suppliers, logistics service providers, manufacturers, distributors and customers on different hierarchical levels. Other (typically software-based) tools available are e.g. ARIS-Toolset (see: http://www.aris.com) or Unified Modeling Language (UML)-activity diagrams (see: http://www.uml.org), which also rely on generic modelling standards and thereby support the original mapping, analysis and purposeful redesign of processes. However, no matter which tool is in use, the typical output are process-flowcharts, visualising the respective interrelationships from diverse perspectives (e.g. organisation-, process-, information perspective). Regardless, which concrete reference model is applied, the reasonably detailed information about the current process landscape in place is the cornerstone for sustainable improvements on the way to best logistics practice.

Case:
IKEA – Reconfiguration of the Supply Chain Structure
One of the main learnings here was that the reconfiguration of the supply chain structure is basically a *process oriented endeavour* which has huge influence on other supply chain areas, i.e. transport, warehousing, procurement, and distribution.

(See Case Collection in Part IV of this book)

4.2.2 Analyse and Monitor Systematically

The next step on the way to sustainable process excellence consists of a systematic analysis of the mapped processes in order to find out how your processes perform individually and how the system of interrelated processes performs as a whole. By using an appropriate set of quantitative (monetary, e.g. transportation costs, and non monetary, e.g. lead times) *key performance indicators* (KPI) input, throughput, and output indicators are measured and monitored on a regular basis. Here, next to other means, the toolbox of the SCOR-Model also offers predefined KPI-metrics for performance measurement. In general, such an analysis enables best practice companies to understand the value contribution of each internal and external business process involved, as well as their value tradeoffs systematically. Furthermore, this systematic analysis reveals the impact of critical processes on other processes' performance and the performance of the whole process system. Especially with regard to sustainable SCM the best practice companies do not only monitor and analyse their economical performance. Additionally, they accurately shape a system of social and environmental KPIs. Based on the popular "balanced scorecard" a yet emerging tool applicable here is the Sustainable Supply Chain Scorecard (SSC)-Scorecard (more about SSC-Scorecard was presented in Chap. 2), which aims at the integration of economical, social and environmental goals by combining respective KPIs on a process perspective.

Case:
Shell Chemicals Europe and Bertschi AG
One of the main learnings here was that the joint definition and monitoring of KPIs and the quick reaction to their developments is a critical success factor.
(See Case Collection in Part IV of this book)

4.2.3 Improve Continuously

Knowing (i.e. mapping) and understanding (i.e. analysing and measuring) your processes are essential prerequisites for improvement on the way to sustainable process excellence. Hence, best practice companies are never satisfied with the status quo, but continuously question their current practices and consequently strive for change, if e.g. environmental, technical or competitive factors have changed. The well known, process based philosophy of KAIZEN, originated in the concept of the Toyota Production System is an example for this strive towards process excellence. Another prominent method applicable here is benchmarking. Usually, in search for process improvement, the identified processes and their performance are compared with corresponding measures from other organisational units and/or other companies depending on the benchmarking approach (internal/external benchmarking).

Identified performance gaps are then analysed in depth in order to find out the relevant areas for process improvement. Certainly, best practice companies strive for process improvements with regard to economical, social, and environmental goals, simultaneously. These improvements are ideally motivated by a high involvement of the affected employees. Therefore, the involvement of self-motivated and creative employees requires concerted trainings and an overall spirit of leadership dedicated to people. We will focus on these aspects later on in this chapter.

Case:
IKEA – Air Hunting Competition
One of the main learnings here was that small but continuous improvements will yield better overall improvements and cost benefits than a small number of large projects that are more likely to fail to achieve their goals.
(See Case Collection in Part IV of this book)

4.2.4 Know Your Customers

Another critical prerequisite for reaching sustainable process excellence in a SCM-context is a thorough understanding of your respective customer's needs. By applying the approach of process management, best practice companies have figured out, that next to the final (external) customer, which are typically in the primary focus, there are many intermediary (internal) customers to be served by every process in place. In that sense – the original supply chain is rather a chain of customers or a demand chain in which the products are pulled through each process. Obviously, organisations are more and more watched closely, regarding their efforts for sustainability. Nevertheless, companies can promote sustainability efforts of their partners and eventually, the final customer. Therefore, in order to design and improve processes, you have to define your customers thoroughly and to understand the respective requirements of your multiple customers – internally and externally. Only if all internal customers are served with high quality, the final customer will be served with high quality as well.

4.2.5 Segment Your Supply Chain

Companies showing sustainable process excellence reveal the fundamental understanding that one single supply chain process cannot fit to all business models or competitive strategies. Therefore, these best practice companies execute a strategic process alignment by segmenting the supply chain according to consistent combinations or bundles of product characteristics and customer requirements.

For example, with regard to the kind of product and its demand pattern supply chains are segmented into responsive or agile supply chains for innovative products (e.g. fashion apparel) and cost efficient or lean supply chains for functional products (e.g. milk). Lean supply chains are cost-oriented to economies of scale in production and transportation in order to provide reasonable prices. The design of agile supply chains needs to be able to quickly react to unpredictable demand variations. Therefore, agile supply chains typically use fast (and mostly more pollutive) means of transport in order to keep the risk of stock outs or overstocks to a minimum. Hence, creating sustainable supply chain management process in agile supply chains poses major challenges, since agile supply chains ask for quick and thereby often environmentally questionable means of transport. In order to reduce the environmental burden of agile supply chains, best practice companies apply the latest technologies such as new generation trucks, telematics or RFID to enhance energy efficiency and information transparency in the supply chain. Apart from technological improvements the "re-invention" of higher stock-keeping and better forecasting-methods lead to less pollutive supply strategies and the according means of transportation.

4.2.6 Enhance Efficiency and Minimize Environmental Impact

As mentioned before, traditional best practices concepts of process management originally exclusively referred to economical benefits of process improvements. However, nowadays this has changed, since best practice companies additionally consider the environmental and social impacts of their sustainable process solutions. For the advanced trailblazers, sustainability, which is implemented in a serious way in the configuration of their processes, initially offers the opportunity to distinguish themselves within the market. We found out that customers gladly use environmental transports – as long as this specific transport is not more expensive than the original, more polluting one. Hence, viewed over the long term, sustainability will move from being a distinguishing feature to the standard, so that it is inevitable that those protagonists that today adopt a passive or defensive stance will have to fear for their existence in the long term.

Case:
Cargo Domizil – Intermodal Less Than Truckload Transport
One of the main success factors here was that custo-mers preferred "green" transport to conventional road haulage, but only when offered at the same price.

(See Case Collection in Part IV of this book)

4.3 Managing Products

Products represent the major output and the "raison d'être" of any market oriented company. By the kinds of products offered to different target customers and markets, a company decisively determines its branch of business, the associated resource requirements, as well as the originalities of the competitive environment and its respective profitability. In this regard, with the term "product" we denote both, i.e. tangible (physical) products (e.g. cars) and intangible products, commonly labelled as services (e.g. consultancy). The capability to purposively invent, design and efficiently provide products according to the customers desires is key to sustainable profitability of any company. In addition, especially in a supply chain environment of interlinked companies the characteristics of a product, e.g. its functionality, compatibility, handling or packaging, etc., can heavily influence the overall process performance of the whole supply chain.

Managing products generally deals with the respective tasks of intentionally inventing, designing and providing the appropriate product range of a company. In the context of sustainable SCM, managing products encompasses the systematic application of specific methods, tools, techniques and systems in order to ensure an economically, environmentally and socially efficient flow of raw materials, semi fished goods and final products through the supply chain. By this means, aspects of purchasing, manufacturing, distribution and logistics have already to be taken into consideration at an early stages of product design. Furthermore, the individuality, as well as the variety of the range of products has to be monitored and adjusted continuously.

As a consequence, managing products in a well considered and systematic way is an eminently strong enabler for the establishment, maintenance, and improvement of sustainable business concepts in logistics and supply chain management. Against this background and in view of the multiple expert discussions as well as the bestLog case collection, the following four guidelines for "managing products" can be derived, which basically reflect the experiences of those companies showing a best logistics practice in the sense of bestLog.

4.3.1 Adjust Your Product Customisation

Offering suitable and appealing products is a critical task for any company, since it is crucial for survival, and most customers willingly and frivolously state highly individual desires. As a consequence, many companies are very vulnerable to over-individualise and over-engineer their products in order to offer a broad variety of products without considering the possible (negative) impacts to the supply chain. However, best practice companies show, that the degree of product customization can be the result of a deliberate balancing process between definitely required individualisation, possible product standardisation, confined process design and supply chain performance. For example the commonly known and rather simple ABC analysis could be used as a guideline to explore the necessary degree of product customization vs. standardization. Products from category "A" – i.e. low volume proportion of about 10–20% and a high value proportion about 60–80% from total – usually need a high degree of customisation by individual product features enhancing diversity. By contrast, "C"-products show a rather high potential for product standardisation and low diversity. Though defining product customisation and respective product diversity is a very company individual task and it is therefore hard to give an commonly applicable advice, the general guideline for best practice should be: Thoroughly adjust the necessary product customisation in order to keep product diversity to a manageable level.

4.3.2 Estimate and Actively Manage Product Life Cycles

Apart from some exceptions, no product lives forever and products typically show some sort of life cycle in terms of the phases of *introduction*, *growth*, *consolidation*, *decline* and *removing* from the market. Each of these phases exhibit different performance requirements of the product to logistics and SCM. Hence, best practice companies systematically analyse and actively manage their product portfolio in terms of the respective product life cycles in order to align adequate logistics operations throughout the supply chain. For example, products revealing characteristics of the growth phase, are apparently needed in higher quantities, than products in the declining phase. Furthermore, the estimation of products according to its life cycle helps these companies to identify strategic gaps in the product portfolio, needing a replacement. As a consequence best practice companies systematically integrate the life cycle view into the management of their product portfolio.

4.3.3 Establish an Integral "Design for Sustainable SCM-Concept"

During the product design process many characteristics of a product are defined, which frequently cannot easily be modified later on. Hence, in order to avoid costly design changes during the product life cycle, it is crucial to anticipate as many possible requirements with regard to e.g. manufacturing/assembly, logistics, packaging or disassembling/recycling. Best practice companies, therefore know the diverse "Design for '...'-concepts in place" (e.g. "Design for Manufacturing", "Design for Assembly", "Design for Disassembly/Recycling", "Design for Logistics") and smartly combine them in order to an integral "Design for Sustainable SCM-Concept". Aspects of sustainability should therefore pervade all design activities affecting the product during its life time – starting at the selection of the right raw materials, social, economical and environmental aspects of production, packaging and transportation, ending at the redistribution, disassembly and recycling of the product. Best practice companies nowadays typically approach this topic by establishing an integrated calculation of a "carbon footprint". Needless to say that this is a very high demanding task which has to integrate nearly all functions of the supply chain.

Case:
Inditex Pro-Kyoto Project
One part of the "Kyoto-Pro" project of Inditex is the measurement of the carbon footprint of the relevant processes in the supply chain.
(See Case Collection in Part IV of this book)

4.3.4 Control Parts Variety

Next to the variety of final products, the variety of parts plays a further major role in coping with complexity and supply chain performance, respectively. Best practice companies strive for control here, because rising parts variety increases complexity and cost. Controlling parts variety encompasses all those means for organising a parts pyramid of a product assembled from many similar parts each comprising a combination of a multiplicity of component parts. One prominent concept here is modularity of parts and components, e.g. applied in the mass-customisation production. Hence, in their general search for simplicity, best practice companies carefully evaluate and control the impact of rising or decreasing parts variety. The concept is also applicable to services or service modules which can be attached to the products (e.g. installation services) or can be a product on their own (e.g. in logistics service providing).

4.4 Managing Knowledge

Nowadays knowledge has become a key factor for the sustainable competitiveness of companies. Every innovation in an economical, ecological or social way is only possible by using and creating specific knowledge. Hereby knowledge is not a classical tangible factor of production, but rather inherent in people and thus intangible. The term knowledge originally means the linking and decoding of specific information by people. Hence, it can be derived that each person not only has individual knowledge, but different kinds of knowledge. Especially in the post-industrial society, characterized by globalization, augmented competition and highly sophisticated technologies, the fields of knowledge became and still become more and more differentiated and complex. This sets new challenges for companies in the future. An organization meeting these emerging demands in an adequate manner needs to possess differentiated knowledge on processes, products and infrastructure. As a logical consequence of the risen complexity and comprehensiveness of knowledge, knowledge is usually embodied in generalists on the one hand and in specialists on the other hand. This implies two different kinds of depth of knowledge – either bound in generalists or in specialists. Generalists deal with comprehensive knowledge and are able to separate the knowledge according to the degree of importance. Specialists are able to apply in-depth knowledge and thereby typically enhance the knowledge base.

The growing complexity of knowledge as well as the according importance for the continuous existence of a company requires processes for professional dealing with knowledge in terms of the so called knowledge management. Knowledge management is a term for all operations and management tasks, aiming at an optimal dealing with knowledge in terms of targeted development of knowledge. Knowledge management hereby intends to provide specific tools to collect, save, systematically prepare and eventually disseminate knowledge throughout the organization and the supply chain. Therefore, in order to maintain and enhance knowledge in a company it has to be kept "alive". Hence, knowledge management aims on fostering and offering an incentive for transferring know-how to spread relevant knowledge within a company and motivate people to share their learnings and experiences with their corporate colleagues as well as the relevant supply chain members.

Starting from the internal perspective on passing on and spreading of know-how, companies can extend these proceedings to the whole supply chain in order to receive enhancements and sustainability thinking with all their supply chain partners. Knowledge on sustainability driving aspects, like e.g. process optimization, applied in the entire supply chain lead to more effective and competitive supply chains. Sharing knowledge needs enhanced communication between the supply chain partners, which provides the opportunity to use these communication channels to act in a concerted way through the whole supply chain and hereby improves efficiency and hence sustainability. Sustainability in terms of social aspects are additionally met by an enhanced skill level of the employees due to augmented knowledge in the supply chain, leading to sound satisfaction of the employees and enhanced chances for a personal continuous development.

Consequently, managing knowledge in an integral and progressive way is a fundamental enabler for the sustainable businesses in the supply chain management context. Against this background and the view of numerous expert discussions as well as the bestLog case collection, the succeeding five guidelines for "managing knowledge" have been derived, which basically reflect the experiences of those companies showing a best logistics practice in the sense of bestLog.

4.4.1 Define and Develop Relevant Fields of Knowledge

The initial step on the way to managing knowledge with a sustainability background is to understand *what* will be critical to *know* in the future. Generally, the knowledge of the world is growing rapidly given the fact that the amount of knowledge doubles within increasing shorter time periods, generating increased complexity of knowledge. In order to cope with the risen complexity of knowledge, best practice companies strive for a clear definition and development of relevant fields of knowledge. These fields are often strongly interlinked to or influenced by the company's individual or supply chain's strategy, as strategy describes the roadmap to the company's future. According to the strategy the relevant fields of knowledge have to be developed. Here, best practice companies usually rather define their strategy first and then map their existing knowledge. Nevertheless, a strong interrelation between the existing knowledge and the fields to be developed can be assumed. These existing knowledge fields are then developed according to the aim of an organisation.

Case:
Telematics at SMEs
Sieber decided to use progressive technologies for their truck fleet. So together with a professional provider of telematic systems they developed knowledge in this specific field and can now save on fuel and enhance customer satisfaction by further IT-functions. Additionally they enhanced their employees' motivation as well as level of education
(See Case Collection in Part IV of this book)

If an organisation strives for an enhanced performance in sustainable supply chain management, relevant fields of knowledge have been identified in the following areas:

- Leadership and communication
- Technology
- Process optimization

Methods of leadership and communication describe a highly relevant field of knowledge as it is important to effectively anchor sustainability thinking into the staffs' mind. Communication skills and know-how hereby serve as a means of facilitating the anchoring not only within ones company but throughout the entire supply chain.

Technologies as an essential field of knowledge for sustainability orientation in a supply chain context are characterized by e.g. efficiency driven enhancements and the use of less polluting means of production. For example the use of telematics not only provides better information on the truck fleet for the dispatcher but also animates the truck drivers to save on gas and lorry wear out. Additionally, in terms of the social aspect of sustainability, telematics related systems ameliorate safety for the drivers for example by radar electronics securing the right space to the truck driving in front.

Know-how on process optimization has shown to be a vitally field of knowledge as it enables efficient structures and procedures saving on time, energy and money (Please see Sect. 4.2).

More detailed knowledge fields can exemplary be identified in form of generating knowledge for example on external stakeholder integration, cross-functional integration or cultural framing. External stakeholder integration allows to access external knowledge and credibility needed for a comprehensive design of sustainability initiatives. Cross-functional integration blends the different organizational functions and helps to design feasible, process-oriented sustainability initiatives. Cultural framing enables companies to question business practices and communicate the sustainability initiative's benefits to influential stakeholder segments.

Nevertheless, defining the relevant fields of knowledge hereby is highly individual for each company and usually a strategic task. Supply chain managers may support the general management by providing the according information.

4.4.2 Transform Tacit Knowledge into Explicit Knowledge

On the search for relevant knowledge fields one has to bear in mind, that there are two types of knowledge to differentiate. One type is the so called explicit knowledge and the other one is called tacit knowledge. Explicit knowledge, on the one hand is an evidently existing knowledge which can be codified, transmitted to and shared with other individuals. Differing from explicit knowledge, tacit knowledge on the other hand is rather a kind of knowledge, which people have but consciously do not know, i.e. they are not able to articulate it in formal language and communicate it to other individuals or write it down – *they just know*. As tacit knowledge is usually characterized by personal experiences or reflected by intuition, it is obviously difficult to share this kind of knowledge. Nevertheless, especially this form of knowledge usually is of high value for companies. In order to foster the sharing of this knowledge, understanding the process of knowledge creation is important. Knowledge creation and development *between* people is usually done, when people articulate what they think and clarify their "view of the world". This process of transforming tacit knowledge into explicit knowledge is known as codification or articulation. Best practice companies stand out for being particularly able to transform tacit into explicit knowledge as they are able to enhance the knowledge flow through the whole supply chain.

Techniques to support the process of capturing tacit knowledge are for example e-learning tools. E-learning comprises the use of digital media for presentation and distribution of learning material in order to support communication between employees. A rudimentary and quite affordable mode of transferring tacit knowledge to explicit knowledge and additionally making it accessible for other members of the organization are database-based knowledge platforms. The scheme of these databases works fairly easy as each employee is able to post information on specific topics on the homepage or the company wide intranet. Other members of the organisation can easily access the information and additionally contact the author to gain deeper knowledge in face-to-face interaction. There are further divers professional knowledge management tools available on the market like "Tacit software", "hummingbird", "backweb" and many more. Each company here needs to identify their knowledge management system according to their specific needs and their branch.

4.4.3 Motivate Communication

As knowledge is highly linked to people and their way of thinking it cannot easily be "copied and forwarded". Rather interaction between employees and supply chain members is necessary to forward and also generate knowledge. In order to motivate this interaction, communication is crucial. The kind of communication and the willingness to share knowledge is highly linked to the culture of a company and in the supply chain context linked to the trust between supply chain partners and the possibilities for the employees to interact with each other. Communication and the trustful intercourse is a major lever for implementing sustainability programmes and concepts through the entire supply chain. Here management and especially supply chain managers play a major role as they are part of the external contact of a company. Hence, the management is responsible for an open culture of fruitful communication and exchange within and outside an organisation. Motivation for communication is important in order to keep specific knowledge in a company or a supply chain alive and usable. This avoids the loss of knowledge when employees leave the company to work with someone else or retire. On a supply chain scale maintaining knowledge means being able to fill the gap and not loose know-how if supply chain partners are switched. Best practice companies enhance communication by cross-organisational events and information flow through the supply chain. Informal interaction an communication between employees, supported by informal events, has proven to be a good native soil for conjoint projects for sustainability.

Case:
Telematics at SMEs
When Sieber implemented their telematics system, it was important to ensure the employees commitment and understanding. Hence, special trainers communicate their know-how on efficient driving all the time to the drivers, assuring enhanced knowledge for the drivers. This procedure was intentionally launched by the management to motivate communication and thereby diffusing knowledge.

(See Case Collection in Part IV of this book)

4.4.4 Establish a Culture of Lifelong Learning

Apart from sharing knowledge, the generation and development of new knowledge is essential for successful companies. Here it is particularly important to establish a culture of lifelong learning. This means that each employee keeps his knowledge up to date no matter at what age he is. On the one hand, this is an important benefit for the company as new knowledge is created or carried into the company. On the other hand, lifelong learning is crucial for the qualification and the future chances of employees. Additionally creativity is encouraged by a continuous learning and also contributes to the employee's satisfaction as he actually fells like a irreplaceable member of the organisation. Especially when talking about the social side of sustainability, job satisfaction reflects a major goal of sustainable enterprises. For example, lifelong learning can be supported by providing the opportunity to get access to training- and educational programs according to the employee's interest and job focus. Especially as sustainability know how is growing more and more important, education with respect to the three dimensions (economic, ecological and social) offers chances to establish a culture of lifelong learning with sustainability background. A wider focus as well as alternate and defiant challenges are established by implementing job rotation mechanisms, where employees switch jobs with their colleagues for a specific term in order to understand his work and possibly create new solutions by transferring their experiences to the "new" job. By this, innovation is fostered due to transferring know-how from one division to another.

Here also the identified relevant knowledge fields are reflected in the process of lifelong learning. The culture of learning and communication offers the opportunity for elder employees to pass on their knowledge on the experiential driven fields of leadership and effective communication processes while younger employees can share the knowledge on new technologies and process improvement approaches (Please see Sect. 4.2).

4.4.5 Enable Information Transparency by IT-Tools

As mentioned before, the passing on of information is crucial for dealing with knowledge in companies as well as in supply chains. Evidently it is impossible, that each employee passes on his knowledge to every other employee. The long distances which are the basis of the worldwide division of labour and global supply chain management, are a familiar obstacle for the sharing of knowledge. In order to cope with this challenge, best practice companies use technological instruments, especially IT-tools, as those serve as a mode of spreading knowledge to areal dislocated employees and throughout the supply chain. In particular against the background of the need to enhance sustainability, travelling can be reduced and natural resources thereby be saved.

IT-tools can especially serve as means of spreading information and knowledge to spatial dislocated parts of the company. Particularly if an organization already dealt with a certain topic in a specific region, the learnings could be useful for other subsidiaries or supply chain partners to avoid the repetition of a time consuming learning process. IT hereby helps to save, systematically structure and provide information on specific processes and proceedings and spread this information in real-time throughout the supply chain. Employees and interested parties can independently acquire access to databases and enhance their knowledge. This augments effectiveness and reduces time consumption for developing knowledge twice.

As examples for IT-tools that enhance information transparency video conferences, document management systems and data warehouse systems can be named.

4.5 Managing Infrastructure

From a supply chain management's point of view, managing infrastructure comprises much more than connecting buildings to roads, highways, water and energy supply. Managing infrastructure, as a lever of managing inside your organisation, has to ensure the construction and maintenance of all property, plant and equipment which contribute to fulfilling a company's objectives and hereby cope with current economic, ecological and social requirements. Thus, the management of infrastructure spans all tasks of designing, planning, maintenance, controlling and monitoring of facilities and information systems. It denotes the implementation of methods and instruments to ensure the availability of all physical resources, besides raw materials and other supplies. Objects of infrastructure management are usually characterized by a long-term character. Hence, managing infrastructure has a strong component of anticipating future developments and according needs for infrastructure.

Especially when dealing with supply chain management, infrastructure plays a major role for prevailing sustainability issues. To give an example one could instance the working conditions defining character of infrastructures, as the buildings etc. used in supply chains have a major influence on the employees' safety and health as well as job satisfaction. As an outcome of bestLog, especially planning of locations and facilities has to take environmental issues into consideration. In a supply chain context, environmental effects can be reduced by a careful selection of the adequate location and the respective building. The availability of external infrastructure or proximity to suppliers and customers directly impact environmental effects. As the requirements in the business environment and accordingly in supply chains varies constantly, a major aim of managing infrastructure is, according to bestLog results, the assurance of flexibility for the usage of infrastructure; i.e. that space can be used for multiple purposes. E.g. a warehouse is constructed and built in the way that allows to store any product, independent from size and weight. As no special space for a single product needs to be reserved and hold available, the space can be used in multipurpose, leading to smaller warehouses and less need for land and less sealing of the soil surface.

Hence concluding, managing infrastructure in a sound and structured way is a clear and long-term lever for sustainability in logistics and SCM. In this regard and in the light of the multiple expert discussions, as well as the learnings from the bestLog case collection the following guidelines for managing infrastructure were derived, which again reflect the experiences of those companies showing a best logistics practice.

4.5.1 Map and Improve Your Location in the Supply Chain Network

Supply chain networks consist of a system of companies respectively organisations involved in the value-added process. The network ranges from the supplier's supplier to the customer's customer. Mapping the own position in this supply chain network is crucial to evaluate the current strategic position within the supply chain. The visualization of supply chain networks by the aid of modelling instruments supports organizations in sketching out upstream and downstream processes. An allocation of infrastructural characteristics can improve not only the own position, but the position of suppliers and customers in the supply chain network, too.

Case:
Hamé Distribution Centre
Hamé mapped its location in the supply chain network and built a new customized distribution facility resulting in less kilometres driven, damaged goods rate, error rate, and time the goods stay in the DC
(See Case Collection in Part IV of this book)

Infrastructure management also deals with classic tasks of supply chain network design, where each facility role is determined by defining all procedures in each location. Based on this, the facility locations are determined. When specifying facility locations one needs to bear in mind, which products and which capacities are provided by each facility. Thesefore, each company has to identify, what kind of market at which location should be served and by which facility. Additionally, the supply sources for each facility are determined and located. Best practice companies showed, that these principal considerations for building a supply chain are run through over and over again, questioning constantly the actual location of each supply chain member. This constant process allows to enhance the logistics network with small steps rather than fundamental reorganization as a reaction to changed circumstances in the business environment.

4.5.2 Level Your Capacities to Supply and Demand

Subsequently to mapping one's location in the supply chain network, best practice companies level their capacities to supply and demand in order to avoid unused capacities and infrastructure resulting in poor efficiency and hence sustainability performance.

The degree of utilization of existing capacities is above all determined by the demand side on the one hand and the supply side on the other hand. Capacities should always be oriented to the demand in order to keep the stocks on a low level and produce the right product for the right customer at the right time and the right amount. But satisfying the demand with the according products can only be met, if the right supply is assured. Utilization of capacity then is the bound between supply and demand and essentially influenced by these two dimensions.

Case:
Consolidation of Transports
The customer of the LSP Sieber required additional capacities for warehousing due to risen demand. These capacities could either be realized by an own, new warehouse or sourced out. In order to have the best levelled capacity to the supply and demand Sieber provided an adaptable warehouse leading to less costs for the costumer and additionally avoiding construction costs for the new warehouse. Information sharing proved to be crucial for this relationship

(See Case Collection in Part IV of this book)

Especially when dealing with sustainable supply chain management, the right capacities at the right time at the right place are fundamental for efficiency and reducing waste.

Levers for influencing the capacity according to the demand are long-term and precise demand forecasts, as they are essential for an effective management of capacities and hence infrastructure. As for best practice companies this action is accompanied by active and agile sales teams to anticipate the demand for the future by being able to listen to the needs of the market carefully. Dependable forecast allows to adjust the capacities and the supply. Adjusting supply is mainly done by allowing quick reactions of the supplier base through instant information sharing. Allowing an intense information flow between supply chain partners requires close relationships between your company and the suppliers.

Infrastructure management then needs to take care of the respective information and technique equipment for sharing the information through the supply chain. Best practice companies show to be capable of providing modern and flexible IT-infrastructures. As there are usually various IT-systems in use throughout the supply chain, these IT-infrastructures need to be easy adoptable to the supply chain partners IT in order to allow a barrier free information flow.

Apart from the outside-defined dimensions demand and supply, best practice companies are able to provide flexible structures in capacities. Flexible capacities can for example be realized by implementing flexible working conditions for the employees. We deal with this in Sect. 4.5.4 of this chapter.

4.5.3 *Design and Maintain Your Infrastructure According to Ecological Standards*

As mentioned above physical infrastructure plays a major role for improvements in terms of sustainability issues. Especially when building new infrastructure, sustainability concerns can easily be taken into account as widely recognized ecological standards and enhancements for infrastructure have been made available in the past. But not only the new design of infrastructure holds the opportunity to improve sustainability in supply chain sites. Adjusting all internal infrastructures to recent ecological standards or the usage of renewable energy seems expensive on the short run, but usually reduces energy cost in long-term. Moreover, governmental legislation influences infrastructure management introducing technical and ecological standards. A proactive management of infrastructure in terms of ecological and technical standards can then form a competitive advantage. The example of chemical industry illustrates this point: to fulfil governmental standards, extremely high capital investments in filter systems for industrial wastewater are required. By proactively implementing filters, companies were able to outperform competitors that could not adopt the technology fast enough. By the time using filter systems became obligatory due to a change in legislation the proactively engaged companies already used the standard and did not have to implement it near-term and thereby more expensive as competitors did.

Case:
Migros – Biofuel Trucks
The retailer Migros uses biofuel trucks for the supply of their stores. The required biofuel is produced with organic waste at a nearby plant. This organic waste is delivered by the Migros stores, where the organic waste earlier needed to be disposed of.

(See Case Collection in Part IV of this book)

Other levers in terms of ecological standards for designing infrastructure are for example:

- The avoidance of soil sealing by redirecting the rainwater being gathered at e.g. warehouse roofs is redirected to the surrounding oil instead of the sewer.
- Applying better isolation to safe on energy and heating/air-conditioning costs
- Proactively apply safety instructions for dangerous goods

As pointed out in the example above, designing and maintaining infrastructures according to ecological standards helps sustainability-driven companies to gain a competitive advantage in the market which for instance could use the fulfilment of these standards as a marketing effect. In terms of a supply chain orientation, best practice companies ask for compliance of all supply chain partners to ecological and social standards.

4.5.4 Assure Flexible Usage of Infrastructure

In order to meet sustainability issues, infrastructure must be able to be adopted flexibly to changes in the business environment and demand. Flexible usage allows to reduce over-capacity for decreasing demand and at the same time rapidly satisfy rising demands. Apart from this, especially in a supply chain management context, flexible usage can also be related to the multimodal connection to external infrastructure allowing to transport goods either by road, water or railway. Changing cost structures in road-, rail- and waterway transport can be utilized for the company's benefit and also sustainability performance.

- In general, flexible infrastructure unites attributes like:
- Multifunctional applicable (e.g. use of standardized boxes)
- Mobile (not bound to one location)
- Demand-driven (only allocates the amount that is demanded by the market)
- Multimodality (allows all kinds of transportation concepts)

Case:
Bertschi – Supply Chain Redesign
When the LSP Bertschi took over shells chemical storage they applied a new idea of keeping the flexible storage in form of interchangeable tanks instead of silos. By doing so, LSP could provide the maximum in flexibility at comparably low costs.
(See Case Collection in Part IV of this book)

In order to assure flexible usage of infrastructure companies usually can revert to specific concepts. These concepts can be segmented into three dimensions:

- Technological
- Social
- Procedural

The tree types of concepts will be illustrated by an example for each. Technological concepts for flexible infrastructures are for example the implementation of standardised containers to assure the use for multiple versions of products. A social concept for the flexible use of infrastructures is flexible working schedules, adjusted to the actual demand for the products. Procedural concepts are for example the joint use of one warehouse by two competitors in order to save on fixed costs for both. This requires a change in strategy but enhances the sustainability performance of both companies.

4.6 Managing People and Teams

Companies just as supply chains do not only consist of "inanimate" processes, activities, facilities or organisational units but resemble dynamic and "living" entities composed of people and collaborative team relations. Especially people with their personal talents, beliefs, capabilities and professional know-how represent the main parameters of company or organisational life. This critical aspect of people and teams as "task bearers" who finally execute the diverse activities, who make decisions etc. has too often been neglected by practitioners and management theorists alike. Thereby, management is essentially not only driven by people but also focused on people, just as Mary Parker Follet[1] prominently stated "*Management is the art of getting things done through people*". Hence, people as well as collaborative teams are the decisive enabler but can also be discriminating barriers to managerial change, organisational SC-development and (inter-)company performance. This has to be taken into serious consideration when pursuing and implementing sustainable SCM concepts.

Since the main focus of bestLog is on inspiring ideas for enhancing sustainability in the intra- and intercompany supply chain context and – as a logical consequence – all collected cases more or less show the change of previous (unsatisfactory) practices into new (more satisfactory) practice, it was therefore not surprising for us that the management field "people and teams" has overall been evaluated as a very important management field throughout the industry workshops, the online forum and the case collection, respectively. Additionally, what we have experienced throughout the whole bestLog endeavour is, that different people/ teams play different roles with different stakes on sustainable supply chain performance. Against this background, best practice management of people & teams in the context of sustainable logistics and SCM means a stakeholder aligned management within and between companies.

In general we could identify four basic stakeholder groups as shown in Fig. 4.6. In order to successfully develop and implement sustainable supply chain concepts – or

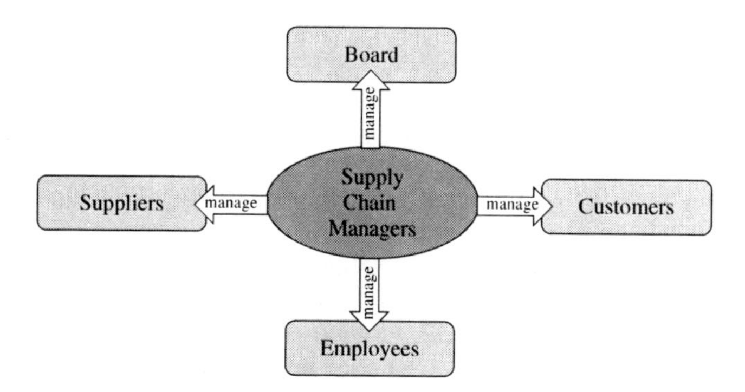

Fig. 4.6 Four Basic Stakeholder Groups in SCM

[1] 1868–1933. Early female pioneer of modern management theory.

in the words of FOLLET "*to get the things done*", the responsible supply chain manager has to be able to manage each stakeholder group in a certain "target group oriented" way. In view of the role-specific requirements, perspectives or political power of each group, the supply chain manager her- or himself plays the role of a "neutral" integration platform, trying to align the different supply chain perspectives as best as he can – employing an overall "birds eye" point of view. The management board at the top of the organisation, for example requires aggregated financial data of supply chain performance upon which important strategic decisions for new investments can be made. Employees doing the operative tasks on the other hand need clear instructions and to get the things done efficiently and to be motivated for innovation at the same time. Beyond the companies boundaries the supply chain manager has to manage supply chain relations to customers as well as to suppliers. It is obvious that this is a very demanding and sensible task, since each group's targets, requirements and power constellations have to be thoroughly understood, the groups have to be involved at the right times and supply chain actions have to be wisely communicated!

Hence concluding, managing people and teams in a structured, participative and sensitive way is a clear enabler for the development and implementation of collectively accepted sustainability concepts in logistics and SCM. In this regard and in the light of the multiple expert discussions, as well as the bestLog case collection the following four guidelines for managing people and teams can be derived, which again reflect the experiences of those companies showing a best logistics practice.

4.6.1 Understand and Speak your Board's Language

The management board of a company represents the highest decision making authority, taking the responsibility for all strategic decisions with high financial and political/cultural impact. Thus, for the successful enforcement of sustainable supply chain concepts and the associated change activities, financial investments, etc. the commitment of the (top) management board is a mandatory and indispensible necessity. Nearly all of our cases reveal this essential need – though not always explicitly mentioned. As a consequence supply chain initiatives have to be promoted to the management board by the best way possible, just to attract the top managements political commitment and financial backup. This also means that supply chain initiatives have to be adequately communicated in the typical language of the top management – i.e. in terms of aggregated financial key performance indicators, investment and payback calculations etc. The intelligent aggregation of key facts is a core characteristic of language here, since top management usually does not have the time and the capacity to deal with all the operational details of a matter. What we learned from many project managers in the context of logistics and SCM is that, in order to successfully develop and implement sustainable supply chain solutions, a supply chain manager has to have the competence to understand and speak the board's language – otherwise the commitment of top management for substantial sustainability initiatives will be rather weak.

4.6.2 *Involve and Consult Your Staff*

In general, the staff enables implementation success but can also be a major barrier to accomplish sustainable solutions in SCM. Since roughly every initiative to sustainable SCM is associated with change in some way or the other (e.g. technical, organisational, spatial change), the concerned employees have to be involved very early in order to catch as broad overall acceptance as possible. In this regard it is crucial to consult the expertise of respective employees not only to gain consensual acceptance (e.g. by the "political leaders") but also to get the best internal information (e.g. of the "functional experts") one can afford for the development of new solutions. The recommendation to involve and consult your staff impressively reflects the two basic rules of life, that you can't change the habits of a lifetime and that you can't reckon without one's host. Thereby, a culture of continuous consultation and involvement of the staff into the ongoing change matters helps to generate a constructive climate of creative innovation, change awareness and implementation support. Last but not least, from a social sustainability point of view, a team oriented involvement and the offensive appreciation of staff initiatives (e.g. suggestion scheme incentives) leads to increased job satisfaction and strengthens the solidarity with the company. At the end of the day it is a question of social sustainability to be able to attract competent and capable staff which have the potential, the will and the power to make your company even more sustainable in the future.

Case:
Sieber – Telematics at SMEs
One of the main challenges here was that implementing new telematic system needed the handling of employees' fears. As employees are often scared of changes, and especially of the monitoring tools provided by telematics, it is crucial to effectively communicate the rationale for the solution.

(See Case Collection in Part IV of this book)

4.6.3 *Team Up with Your Principal Suppliers*

Since suppliers typically are independent companies *outside* your organisation, they are not exposed to your direct managerial control *within* your organisation. But, as we have learned in bestLog and alike is also proposed in the SCM literature, best practice companies team up with their principal suppliers and thereby try to enfold a direct managerial control in the Supply Chain. Thereby, building strong supply chain relationships with major suppliers is a strong enabler to implement inter-company sustainability initiatives in the upstream supply chain. Ideally this can be achieved by intercompany teams, which are composed of experts from both partner companies – forming a quasi-integrated organisation. With regard to enforcing sustainable supply chain initiatives the aims of teaming up with the principal suppliers are twofold here. On the one hand, a cooperative supplier management motivates and encourages the suppliers to act in a sustainable way and to support as well as to contribute to sustainability initiatives. When claiming to be "green" or "social", for example, the compliance to this commitment throughout the whole supply chain must be safeguarded in order to avoid counterproductive effects initiated by a negative public image (e.g. Childwork, tropical deforrestation, etc.). On the other hand, sustainability in supply chains also means not to exploit the suppliers. A long term supplier-partnership promises cost efficient results which will not cannibalise the associated social and environmental goals.

Case:
SIEMENS – e-Supplier Integration
One of the main learnings here was that including suppliers in joint activities is profitable for all parties because sharing information and collaborating improves their performance.

(See Case Collection in Part IV of this book)

4.6.4 *Team Up with Your Principal Customers*

By the same token best practice companies team up with their principal customers just as with their key suppliers. Hitherto, for many companies it has been a basic element of their marketing campaigns to join into customer driven sustainability initiatives. But, what best practice companies show is that sustainability in its core is not a marketing driven topic alone. For best practice companies, teaming up with their principal customers means building strong supply chain relationships in order to enable and enforce inter-company sustainability initiatives in the *downstream* supply chain with a clear customer focus. This extended company-spanning engagement encompasses the joint development of sustainable supply chain solutions in e.g. distribution, transportation, warehousing, product development, IT, etc. Ideally, these intercompany teams enfold the character of quasi-integrated organisations consisting of experts from the partnering companies. Finding joint sustainable solutions in cooperation with the principal customers thereby enhances the chance of rising service quality without restricting the associated economic, social and environmental goals of sustainability. With regard to economic sustainability this joint engagement last but not least leads to better position in terms of competitive advantage, since reliable and sustainable products and services are the building blocks of a stable supply chain relationship with your current customers. Attracting new customers is by far more difficult than developing the present ones.

Case:
Sieber – Consolidation of Transports
One of the main learnings here was that collaboration with the customer is essential in order that both partners understand each other's needs and develop mutual beneficial solutions.
(See Case Collection in Part IV of this book)

4.7 Managerial Challenge: Find and Prioritize your Internal Sustainability Gap

In this chapter we dealt with five areas of internal management for developing sustainability. Additionally we provided diverse levers in each area to enhance sustainability in supply chains by managing primarily inside your organization. As organizations have different stages of maturity concerning their sustainability status in SCM, there are areas where there is still potential for improvement while others are already satisfyingly fulfilled. Additionally, the identification of the major lever is always dependent on the specific situation of a company. Hence, each organization has to find its own prioritization in order to fill the sustainability gap. In the following we will provide a guided gap analysis based on the identified management fields.

The responsible managers need to map their company's current sustainability status and identify the respective gap. The following Table 4.1 will help you to identify the gap. The table is based on the different sections we dealt with before. Those areas as well as the respective levers can be used to first of all identify the gap and finally close it by implementing the mentioned measures. The identification of the company's sustainability status is done by analysing the areas and levers with regard to the actual status of the company.

Additionally, as a tribute to individual contexts and situations, by putting an individual weight on each enabler (depending on industry, company size, company strategy, etc.) managers receive a ranking of their individual internal enablers. Based on this ranking managers may tackle the most urgent ones and hereby improve sustainability from an internal perspective step by step.

The result of this analysis can subsequently be illustrated in a chart like in (Fig. 4.7) (here only exemplary).

This analysis should be undertaken with different stakeholders from different perspectives. The results of this analysis can then be compared with the different stakeholders in order to find different perceptions of certain issues and enablers. Figure 4.8 gives an illustration.

Table 4.1 Gap-analysis for internal enablers

Internal enabler of sustainable SCM	Sustainability performance			Weight	Individual result
	Low	Medium	High		
Managing processes					
Know your Processes!	☐	☐	☐		
Analyse Systematically!	☐	☐	☐		
Improve continuously!	☐	☐	☐		
Understand your Customers!	☐	☐	☐		
Segment your supply chain!	☐	☐	☐		
Enhance process efficiency, but minimize ecological and social impact!	☐	☐	☐		
Managing products					
Adjust your product customization!	☐	☐	☐		
Estimate product life cycles!	☐	☐	☐		
Establish an integral "Design for Sustainable SCM-Concept"!	☐	☐	☐		
Control Parts Variety!	☐	☐	☐		
Managing knowledge					
Define and develop relevant fields on knowledge!	☐	☐	☐		
Transform tacit knowledge to explicit knowledge!	☐	☐	☐		
Motivate communication!	☐	☐	☐		
Establish a culture of life-long learning!	☐	☐	☐		
Enable information transparency by IT-Tools!	☐	☐	☐		
Managing infrastructure					
Map and improve your position in the supply chain network!	☐	☐	☐		
Level your capacities according to supply and demand!	☐	☐	☐		
Design and maintain your infrastructure according to ecological standards!	☐	☐	☐		
Assure flexible usage of infrastructure!	☐	☐	☐		
Managing People and Teams					
Understand and Speak your Board's Language!	☐	☐	☐		
Involve and consult your staff!	☐	☐	☐		
Team up with your principal suppliers!	☐	☐	☐		
Team up with your principal customers!	☐	☐	☐		

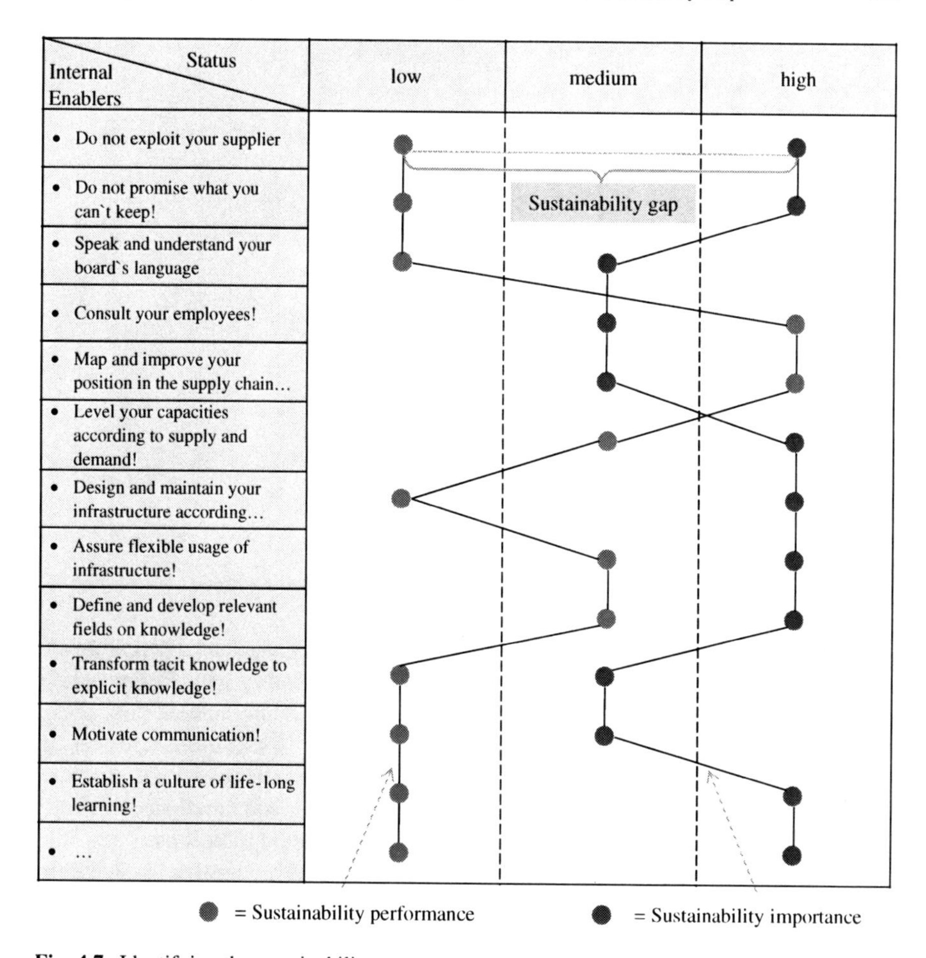

Fig. 4.7 Identifying the sustainability gap

After identifying the internal sustainability gap managers may start with the prioritized improvement of sustainability in their supply chain.

Obviously there is no general checklist or procedure for improving sustainability for each company. In fact each company needs to develop its own measures. The described enablers in this chapter can hereby serve as a guideline.

Generally, in order to analyse the gap supply chain managers need to develop awareness, where internal reasons for a sustainability gaps can arise from. Possible sources may come from conflicts between economical and ecological aspects. Though sustainability aims at balancing and improving each dimension in terms of ecological, economic and social there may still be a conflict of aims when trying to enhance a single dimension. Additionally myopia in management may lead to a sustainability gap, as incentive systems may not reward sustainable thinking; e.g. high variable rewards for managers based on short-run performance indicators abet

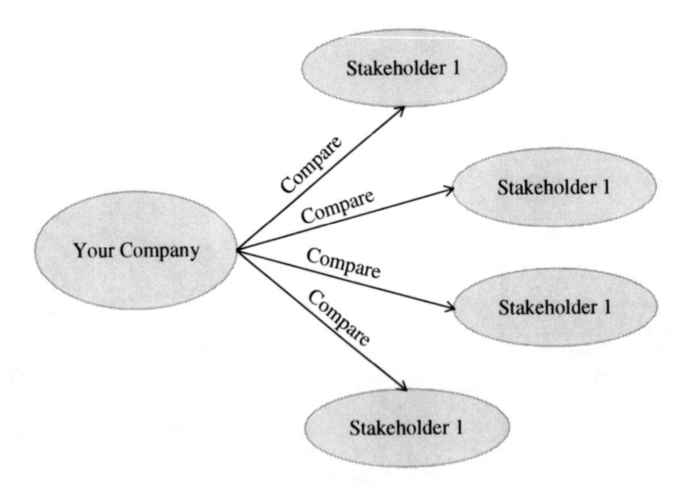

Fig. 4.8 Integrating different stakeholders

adoption of figures quarterly without bearing in mind the long-term survival of the company.

As part of a company's culture missing impulses for changes or sustainable and long-term thinking can be another reason for a sustainability gap. Either certain persons are missing which give the according impulses and implement this special way of thinking or knowledge on sustainability aspects is still quite rare and the awareness that caring about long-term aspects and ecological as well as social concerns are not put on a par with economical concerns. Anchored sustainability thinking hereby offers a major chance to enhance all three dimensions.

Nevertheless, apart from the deductions of this chapter, the bestLog case collection identified a couple of general levers to enhance sustainability. We will deal with those in an exemplary way in the following:

Especially a change in incentive systems turned out to be a major lever for a general orientation towards sustainability within a company. A reasonable way to build the desired mindset towards sustainability aspects is the integration of sustainability measures in the overall performance measurement of managers and a renunciation from the short term orientation of exclusively economical figures.

An advanced system is also suitable for employees. Based on the classic suggestion system, employees can be rewarded for new ideas and suggestions in context of sustainability.

Communicating the importance of sustainability to all stakeholders is essential for the general acceptance.

For covering the costs of implementing such systems a creative search for financial support is helpful. Novel and integral ideas and concepts may for example be financed or subsided by the E.U.

Additionally, valuable inputs come from other companies and organisations. For inspiration and transferring best practices, Part III in this book provides best practice case studies from companies throughout Europe.

4.8 Summary and Conclusions

When dealing with sustainable SCM from the internal view – i.e. within your organisation – multiple enablers were identified. Concentration on processes, products, knowledge, infrastructure and people hereby reflect the main levers for enhancements in terms of sustainable supply chain management. Though promising achievements can be obtained by applying these levers, managers may have to deal with obstacles. As in most businesses "people" is a factor that has to be considered especially as they can on the one hand form a certain kind of blocking for implementing sustainable SCM measures and on the other hand provide the major lever for a successful implementation. In order to achieve the latter, managers have to motivate employees to support the changes due to a stronger sustainability orientation in the supply chain. In order to do so, they have to clarify their decisions and allow transparency. As well as the employees, likewise the board has to understand the measures taken and support them. Here it may especially be important to argue the long term strategy of sustainability actions, as an impact on financial figures may not be directly and in short term visible. The commitment of the top management is also crucial for investments, required for sustainability concepts, as the revenues are often not directly observable (Fig. 4.9).

As exemplified in this chapter, there are certain hints to sustainable supply chain management but there are only few "ready to implement"-solutions to buy in the market. Managers aiming at sustainability actions rather have to create individual

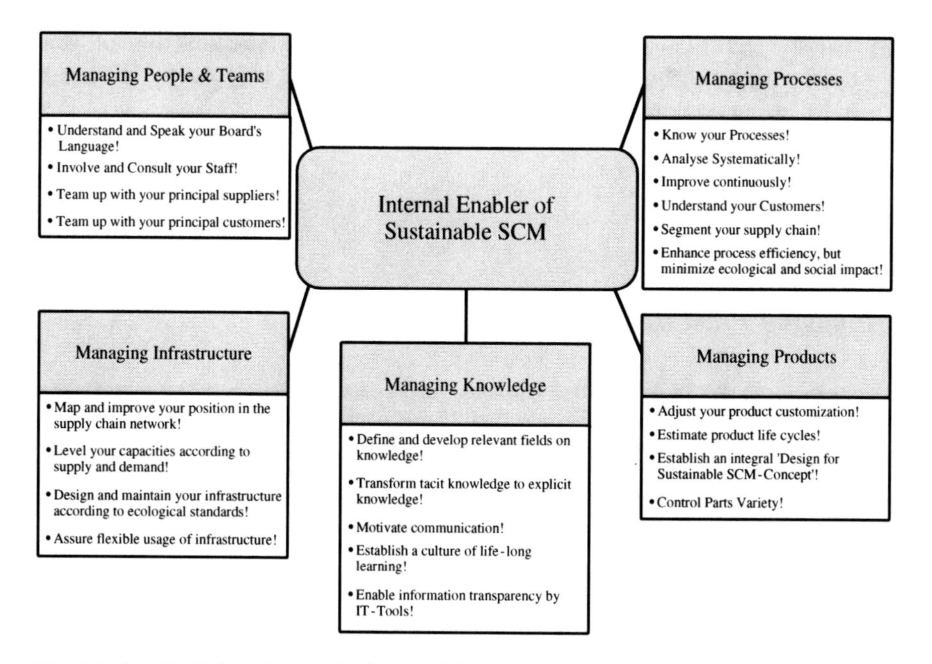

Fig. 4.9 Specific fields of action for internal SCM for sustainability

concepts in order to create new and sustainable solutions for the supply chain. Thereby, it is usually hard to disengage from day-to-day business and take the time to re-think concepts and be creative for sustainable supply chain solutions.

In order to integrate the creative thinking of employees, the installation of a company suggestion plan and the building of trust in the skills and creativeness of employees offers a chance to enhance sustainability in supply chains. Concerning the direct interfaces to the internal management dimensions, managers need sensitise the customers for sustainability issues and integrate them in concepts and sustainable thinking. One major device for integrating and developing enhancements is the use of information sharing systems. Nowadays, technology can provide the distribution of quick and structured information, allowing coordinating and consolidating in order to reduce wastage.

To wrap it up in a nutshell, the use of information systems as well as integration and intended rise of awareness are major levers for enhancing sustainable supply chain management from an organisations inside perspective.

Tasks:

1) Using Table 4.1 rate the sustainability status of your chosen company. Put an individual weight on each enabler to reflect the individual context of your chosen organisation (depending on industry, company size, company strategy, etc.), as well as your individual ranking where appropriate. Transfer the results of your analysis to a chart similar to Figure 4.7.
2) Compare and contrast the analysis in task 1 for different stakeholders in the supply chain.
3) Identify and discuss the priorities for improving the sustainability of your chosen supply chain, given the analyses in tasks 1 and 2, especially with respect to balancing economic, social and environmental goals.

Bibliography

Managing Processes

Childerhouse, P., Towill, S. (2000): Engineering supply chains to meet customer requirements. In: Logistics Information Management, Vol. 13, No. 6, pp. 333–345

Chopra, S., Meindl, P. (2004): Supply Chain Management – Strategy, Planning and Operations. Pearson Education, Inc., Upper Saddle River, New Jersey, 2nd Edition

Croxton, K.L., García-Dastague, S.J., Lambert, D.M., Rogers, D.S. (2001): The supply chain management process. In: International Journal of Logistics Management, Vol. 12, No. 2, pp. 13–36

Fisher, M. (1997): What is the right supply chain for your product? In: Harvard Business Review, Vol. 75, No. 2, pp. 105–116

Handfield, R.B., Nichols, E.L. (2002): Supply Chain Redesign: Transforming Supply Chains into Integrated Value Systems. Prentice Hall, Upper Saddle River

Imai, M. (1986): Kaizen: the key to Japan's competitive success. McGraw-Hill, New York

Kaplan, R.S., Norton, D.P. (1997): Why does business need a balanced scorecard. Journal of Cost Management, Vol. 11, No. 3, pp. 5–11
Llewellyn, N., Armistead, C. (2000): Business process management. Exploring social capital within processes. In: International Journal of Service Industry Management, Vol. 11, No. 3, pp. 225–243
Lovell, A., Saw, R., Stimson, J. (2005): Product value-density: managing diversity through supply chain segmentation. In: The International Journal of Logistic Management, Vol. 16, No. 1, pp. 142–158
McCormack, K., Rauseo, N. (2005): Building an enterprise process view using cognitive mapping. In: Business Process Management Journal, Vol. 11, No. 1, pp. 63–74
Porter, M.E. (1985): Competitive advantage: creating and sustaining superior performance. The Free Press, New York
Sadler, I., Hines, P. (2002): Strategic operations planning process for manufactures with a supply chain focus: concepts and a meat processing application. In: Supply Chain Management: An International Journal, Vol. 7, No. 4. pp. 225–241
Stewart, G. (1997): Supply-chain operations reference model (SCOR): the first cross-industry framework for integrated supply-chain management. In: Logistics Information Management, Vol. 10, No. 2, pp. 62–67
Soliman, F. (1998): Optimum level of process mapping and least cost business process re-engineering. In: International Journal of Operations & Production Management, Vol. 18, No. 9/10, pp. 810–816
Spens, K.M., Bask, A.H. (2002): Developing a framework for supply chain management. In: International Journal of Logistics Management, Vol. 13, No. 1, pp. 73–88
Trkman, P., Štemberger, M.I., Jaklič, J., Groznik, A. (2006): Process approach to supply chain integration. In: Supply Chain Management: An International Journal, Vol. 12, No. 2, pp. 116–128

Managing Products

Ameri, F., Dutta, D. (2005): Product lifecycle management: closing the knowledge loops. In: Computer-Aided Design & Applications, Vol. 2, No. 5, pp. 577–590
Blecker, T., Abdelkafi, N. (2006): Mass customization: state-of-the-art and challanges. In: Blecker, T, Friedrich, G. Mass Customization. Challenges and Solutions, New York, pp. 1–18
Bogue, R (2007): Design for disassembly: a critical twenty-first century discipline. In: Assembly Automation, Vol. 27, No. 4, pp. 285–289
Dowlatshahi, S. (1996): The role of logistics in concurrent engineering. In: International Journal of Production Economics, Vol. 44, pp. 189–199
Fandel, G., Stammen, M. (2003): A general model for extended strategic supply chain management with emphasis on product life cycles including development and recycling. In: International Journal of Production Economics, Vol. 89, pp. 293–308
Forza, C., Salvador, F., Rungtusanatham, M. (2005): Coordinating product design, process design, and supply chain design decisions Part B. Coordinating approaches, tradeoffs, and future research directions. In: Journal of Operations Management. Vol. 23, pp. 319–324
Jacobs, M., Vickery, S.K., Droge, C. (2007): The effects of product modularity on competitive performance. In: International Journal of Operations & Production Management, Vol. 27, No. 10, pp. 1046–1068
Kaipia, R., Holmström, J. (2007): Selecting the right planning approach for a product. In: Supply Chain Management: An International Journal, Vol. 12, No. 1, pp. 3–13
Khan, O., Christopher, M., Burnes, B. (2008): The impact of product design on supply chain risk: a case study. In: International Journal of Physical Distribution & Logistics Management, Vol. 38, No. 5, pp. 412–432

Kotler, P., Armstrong, G. (2007): Principles of Marketing, Prentice Hall, Upper Saddle River, New Jersey, 12th edition
Kriwet, A., Zussman, E., Seliger, G.: Systematic Integration of Design-for-Recycling into Product Design. In: International Journal of Production Economics, Vol. 38, pp. 15–22
Mikkola, J.H. (2003): Modularity, component outsourcing, and inter-firm learning. In: R&D Management, Vol. 33, No. 4, pp. 439–444
Radder, L., Louw, L. (1999): Mass customization and mass production. In: The TQM Magazine, Vol. 11, No. 1, pp. 35–40
Saaksvuori, A., Immonen, A. (2004): Product Lifecycle Management, Springer, New York, NY
Selvaraj, P., Radhakrishnan.P., Adithan, M. (2009): An integrated approach to design for manufacturing and assembly based on reduction of product development time and cost. In: International Journal of Advanced Technologies, Vol. 42, pp. 13–29
Sharifi, H., Ismail, H.S., Reid, I. (2006): Achieving agility in supply chain through simultaneous "design of" and "design for" supply chain. In: Journal of Manufacturing Technology Management, Vol. 17, No. 8, pp. 1078–1098

Managing Knowledge

Boiral, O. (2002): Tacit knowledge and environmental management. In: Long Range Planning, Vol. 35, pp. 291–317
Edwards, R., Ranson, S., Strain, M. (2002): Reflexivity: towards a theory of lifelong learning. In: International Journal of Lifelong Education, Vol. 21, No. 6, pp. 525–536
Fischer, G. (2000): Lifelong learning – more than training. In: Journal of Interactive Learning Research, Vol. 11, No. 3/4, pp. 265–294
Ives, W., Torrey, B., Gordon, C. (1998): Knowledge management: an emerging discipline with a long history. In: Journal of Knowledge Management, Vol. 1, No. 4, pp. 269–275
Liew, C.A. (2008): Strategic integration of knowledge management and customer relationship management. In: Journal of Knowledge Management, Vol. 12, No. 4, pp. 131–146
Maqsood, T., Walker, D., Finegan, A. (2007): Extending the "knowledge advantage": creating learning chains. In: The Learning Organization, Vol. 14, No. 2, pp. 123–141
Neef, D., Siesfeld, G.A., Cefola, J. (1998): The economic impact of knowledge, Butterworth Heinemann
Pitt, M., MacVaugh, J. (2008): Knowledge management for new product development. In: Journal of Knowledge Management, Vol. 12, No. 4, pp. 101–116
Polanyi, M. (1983): The Tacit Dimensio. First published Doubleday & Co, 1966. Reprinted Peter Smith, Gloucester, Mass
Salisbury, M.W. (2003): Putting theory into practice to build knowledge management systems. In: Journal of Knowledge Management, Vol. 7, No. 2, pp. 128–141
Stewart, D., Waddel, D. (2008): Knowledge management: the fundamental component for delivery of quality. In: Total Quality Management, Vol. 19, No. 9, pp. 987–996
Wu, C. (2008): Knowledge creation in a supply chain. In: Supply Chain Management: An International Journal, Vol. 13, No. 3, pp. 241–250

Managing Infrastructure

Braglia, M., Zanoni, S., Zavanella, L. (2005): Layout design in dynamic environments: analytical issues. International Transactions in Operational Research, Vol. 12, pp. 1–19
Chopra, S., Meindl, P. (2004): Supply chain management. Strategic planning and operation. Prentice-Hall, New York

Faisal, M.N., Banwet, D.K., Shankar, R. (2006): Mapping supply chains on risk and customer sensitivity dimensions. In: Industrial Management & Data Systems, Vol. 106, No. 6, pp. 878–895
Gardner, J.T., Cooper, M.C. (2003): Strategic supply chain mapping approaches. In: Journal of Business Logistics, Vol. 24, No. 2, pp. 37–64
Grünwald, C., Gómez, J.M. (2008): Factory planning with regard to environmental information. In: Information Technologies in Environmental Engineering, Vol. 1, p. 23–32

Managing People and Teams

Bullington, K.E., Bullington, S.F. (2005): Stronger supply chain relationships: learning from research on strong families. In: Supply Chain Management, Vol. 10, No. 10, pp. 192–197
Chu, S.-Y., Fang, W.-C. (2006): Exploring the relationships of trust and commitment in supply chain management. In: The Journal of American Academy of Business. Vol. 9, No. 1, pp. 224–228
Gowen, C.R. III, Tallon, W.J. (2002): Enhancing supply chain management practices through human resource management. In: The Journal of Management Development, Vol. 22, No. 1, pp. 32–44
Groves, G., Valsamakis, V. (1998): Supplier–customer relationship and company performance. In: International Journal of Logistics Management, Vol. 9, No. 2, pp. 51–64
Hult, G.T.M., Ketchen, D.J. Jr., Chabowski, B.R. (2005): Leadership, the buying centre and supply chain performance: a study of linked users, buyers and suppliers. In: Industrial Marketing Management, Vol. 36, pp. 393–403
Jeong, J.S., Hong, P. (2007): Customer orientation and performance outcoumes in supply chain management. In: Journal of Enterprise Information Management, Vol. 20, No. 5, pp. 578–594
Koulikoff-Souviron, M., Harrison, A. (2007): The pervasive human resource picture in interdependent supply relationships. In: International Journal of Operations & Production Management, Vol. 27, No. 1, pp. 8–22
Lee, C.W., I.-W. G., Kwon, Severance, D. (2007): Relationship between supply chain performance and degree of linkage among supplier, internal integration, and customer. Supply Chain Management: An International Journal, Vol. 12, No. 6, pp. 444–452
Scarbrough, H. (2000): The HR implications of supply chain relationships. In: Human Resource Management Journal, Vol. 1, No. 1, pp. 5–17
Schneller, E.S., Smeltzer, L.R. (2006): Building supply chain leadership and resources for the future. In: Healthcare Purchasing News, p. 75
Sheth, J.N., Sharma, A. (1997): Supplier relationships. Emerging issues and challenges. In: Industrial Marketing Management, Vol. 26, pp. 91–100
Smyrlis, L. (2005): Human resources: influential study paints a picture of a sector in need of vision and strategic leadership. In: Canadian Transport Logistics, Vol. 11, No. 108, p. 63
Upson, J.W., Ketchen, D.J. Jr., Ireland, R.D. (2007): A key to the effectiveness of strategic supply chain management. In: Organizational Dynamics, Vol. 36, No. 1, pp. 78–92
Zhao, X., Huo, B., Flynn, B.B., Yeung, J.H.Y. (2007): The impact of power and relationship commitment on the integration between manufacturers and customers in a supply chain. In: Journal of Operations Management, Vol. 26, pp. 368–388

Chapter 5
Managing Outside Your Organisation

Balkan Cetinkaya

Learning Goals.
By reading this chapter you will:

- Know who your supply chain external stakeholders are
- Understand their potential impacts on your supply chain
- Understand the action fields for SC stakeholder management
- Understand the need for "dedicated" SC stakeholder management
- Apply effective definitions and classifications of supply chain stakeholders

5.1 Introduction and Structure of the Chapter

Sustainable supply chain management expands the sphere of supply chain strategy and management, adds further metrics to performance measurement systems, and new criteria for operational excellence. Having discussed such issues of strategy, performance, and process excellence in the context of sustainable supply chain management in the last three chapters, we will now shed some light on best practice in the aspects of sustainable supply chain management which involve your organisation's links to the outside world.

These concern areas that are not under your direct control, but where you may exert influence, and where you equally need to be aware of and anticipate the attitudes and actions of other relevant actors – such as suppliers, service providers, customers, governments, local communities, and NGOs.

The following Sect. 5.2 gives you some basic grounding in the topic; (see Fig. 5.1) You will find out who the external supply chain players are and what impact they may have on your supply chain. During the bestLog project, we held many discussions with hands-on practitioners, general industry experts, industry association representatives, academics, and politicians, which we conducted in dedicated workshops, via the bestLog online platform and during case study interviews. From this primary research we identified a number of external action fields,

C. Tyssen et al., *Sustainable Supply Chain Management*,
DOI 10.1007/978-3-642-12023-7_5, © Springer-Verlag Berlin Heidelberg 2011

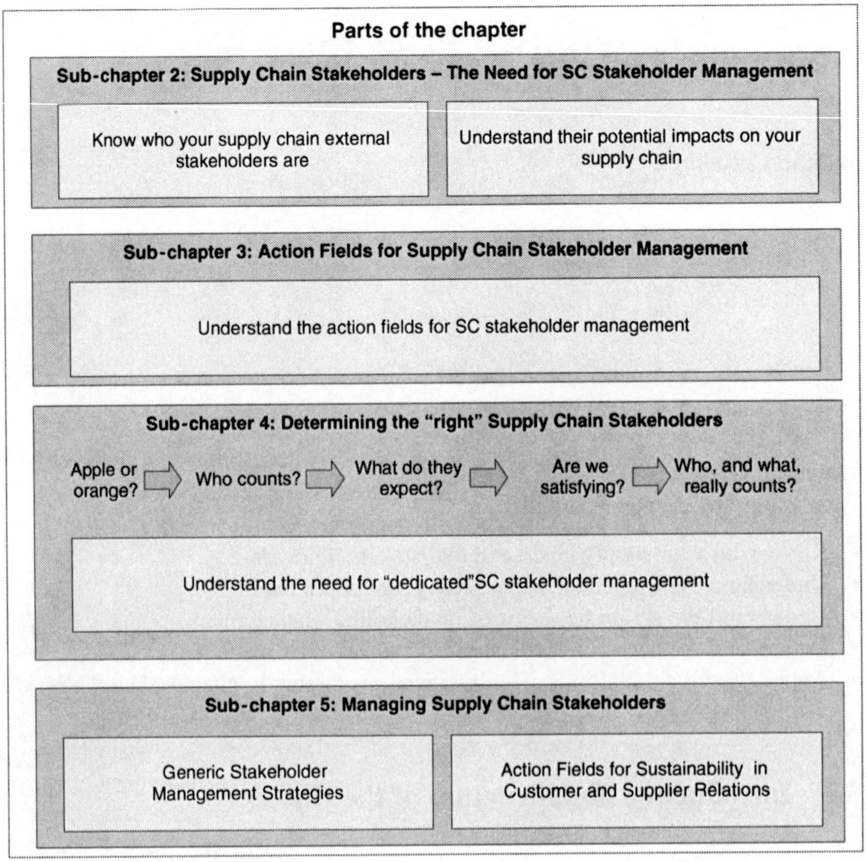

Parts of the chapter

Sub-chapter 2: Supply Chain Stakeholders – The Need for SC Stakeholder Management

Know who your supply chain external stakeholders are

Understand their potential impacts on your supply chain

Sub-chapter 3: Action Fields for Supply Chain Stakeholder Management

Understand the action fields for SC stakeholder management

Sub-chapter 4: Determining the "right" Supply Chain Stakeholders

Apple or orange? → Who counts? → What do they expect? → Are we satisfying? → Who, and what, really counts?

Understand the need for "dedicated"SC stakeholder management

Sub-chapter 5: Managing Supply Chain Stakeholders

Generic Stakeholder Management Strategies

Action Fields for Sustainability in Customer and Supplier Relations

Fig. 5.1 Structure of the chapter

which we will cover briefly in Sect. 5.3. In Sect. 5.4 we will set out a five-step approach to assessing and classifying external supply chain stakeholders. This classification will bring us finally to Sect. 5.5, where we will discuss generic strategies for managing supply chain externals before defining and describing best practice for sustainability in customer and supplier regulations in Sect. 5.6.

5.2 Supply Chain Stakeholders: Who Has Which Impact on Your Supply Chain?

The first question challenged in this chapter is: What parties in the supply chain should be counted as external stakeholders deserving management attention?

The scope definition of who counts beyond the boundaries of your own organisation in the context of supply chain management may be broad or narrow. First, we might ask: Who in the supply chain is important to the core task of a supply chain, namely, supplying? The answer could be as simple as the question: Your customer, because in the end, no orders means no supply. But you would be right to observe that a modern supply chain is not that simple, and defining who is important within it should not be that simple, either. There are many more parties deserving and demanding management attention in the establishment and maintenance of a supply chain, especially a sustainable supply chain; getting it right in today's complex supply chains and networks often represents a competitive advantage for many large companies.

In the 1970s and 1980s, a supply chain manager (assuming that executives with such a title existed then) of a manufacturing company was typically responsible for a supply chain consisting of a manageable number of external suppliers, one or two huge production plants, and a manageable and visible number of customers, and perhaps a few wholesalers or retailers. Today, supply chain managers are responsible for designing, planning, optimising, and running complex supply and demand networks consisting of parties including consumers, third-party logistics suppliers (3PLs), outsourced integrators or "tier-0.5" suppliers (4PLs in Accenture's trademarked coinage), second-tier, n-tier suppliers, remanufacturers, waste management companies, recycling companies, infrastructure operators, Brokers et al.

Taking the broad view of supply chains, there are further groups which play roles to greater or lesser extents relevant to supply chains; see Fig. 5.2 below. According to R. Edward Freeman, the originator of stakeholder theory, all these groups shown in Fig. 5.2 should be called stakeholders. Freeman's general definition of a stakeholder is: "Groups or individuals of an organisation who can affect or are affected by the achievement of the organisation's objectives"[1]. These include, for example: (1) governments, imposing a growing array of international and national regulations; (2) increasingly environmentally aware and demanding customers and other consumers; (3) Closely connected to these consumers are a company's employees, who want to work for an environmentally and socially responsible enterprise; then (4) society in general with its calls for greater corporate social responsibility, voiced increasingly through NGOs' enhanced communications and media reach.

These interests and pressures can significantly affect a supply chain's performance and its value proposition. Legislation may determine, for example, what new logistics infrastructure projects (seaports, airports, warehouses, plants, etc.) may be completed; politics may influence economically but not environmentally optimum location decisions, and decisions on new suppliers and service providers; new regulations may incur rising procurement, logistics or transport cost increases, while industrial relations or environmental pressures may reduce the reliability of logistics, respectively through strikes, or by forcing immature modal shift options on logistics suppliers. The case of Wal-Mart shows just how challenging managing such external stakeholders can be. Wal-Mart, the world's largest retailer, has been put under the spotlight by social activists. The company faces an enormous challenge in balancing low prices with various social concerns. The essence of the

[1]Freeman (1994), p. 46.

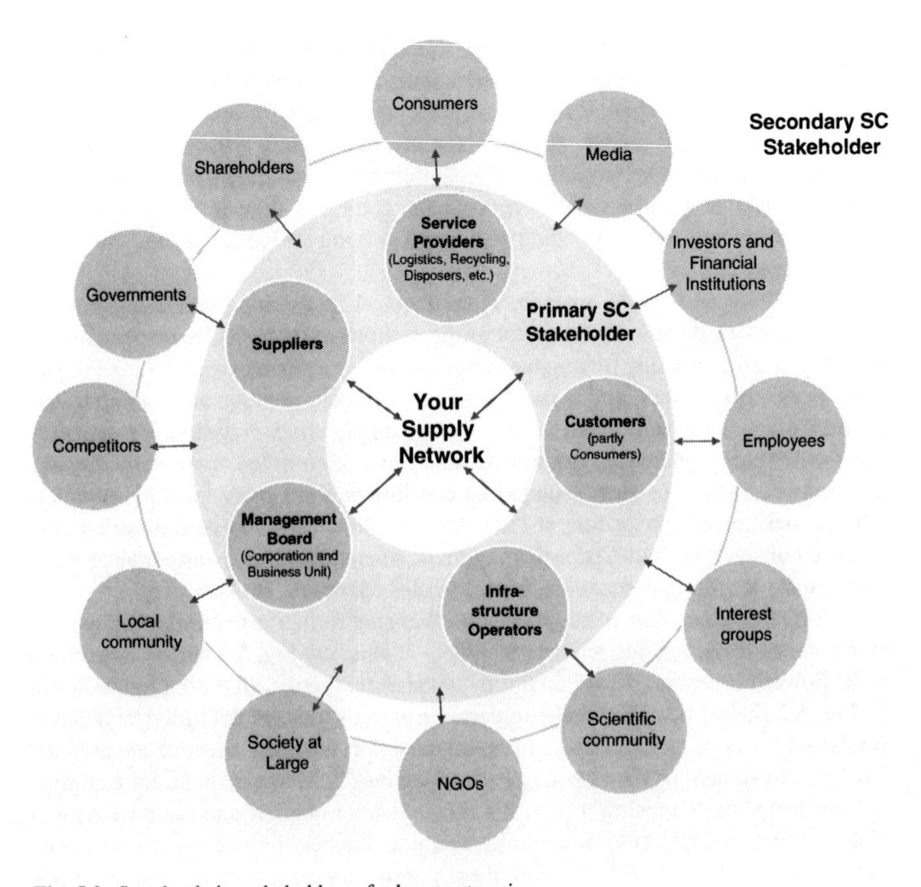

Fig. 5.2 Supply chain stakeholders of a large enterprise

Wal-Mart business strategy is to offer products to the consumer at the lowest price possible. But critics say that Wal-Mart – among others – achieves this through poor conditions for its employees and its suppliers' employees. Wal-Mart's environmental footprint is huge. It is the largest private consumer of electricity in the USA and operates the country's second largest fleet of trucks. It also has the potential to have a substantial impact on sustainability through its supply chain, which includes over 60,000 suppliers. Wal-Mart is making some effort to become more sustainable. However, critics continue to question whether the company will succeed in its plans, and whether its intentions are credible. Some have called its efforts publicity stunts and empty promises.[2]

The Wal-Mart example points to an important fact: not all firms are exposed to the same type of pressure, or to the same extent. Whether the generic stakeholder types we have mentioned are all relevant or deserve to be considered as targets for

[2]Epstein (2008), p. 24.

engagement by a given company's supply chain management can be debated. The definition of key stakeholders will depend particularly on the supply chain impact, the products and services concerned, the industry, the company's size, the geographical region, and other contextual factors such as existing levels of trust, distrust, and image/reputation; stakeholder classifications may be based on hard, measurable facts, but also on perceptions.

In this context, image is a two-edged sword. On the one hand, a positive image helps to generate trust; on the other hand, it entails often substantial reputational risk. Esty quotes a former CEO of Coca-Cola: "When you are the guardian of the world's most valuable brand, environmental mistakes can cost millions, even billion of dollars. This is particularly true for companies with high intangible asset values like Coca-Cola. With a market capitalization of $115 billion, Coca-Cola's book net asset value amounted to only about $15 billion, leaving a whopping $100 billion in intangible value.[3]

In the following brief overview of some relevant supply chain stakeholder groups, we will look at examples of their types of influence. We distinguish, following the narrow and broad views discussed earlier, between primary and secondary stakeholders: Primary stakeholders, shown in the inner circle of Fig. 5.2 are those stakeholders who have formal, official, or contractual relationships with your company, and have direct and necessary economic impacts on the supply chain. Secondary stakeholders, shown in the outer circle, are diverse and include those parties who are not directly engaged in the supply chain's core activities, but are able to exert influence on, or are affected by, the supply chain.[4]

5.2.1 *Secondary, Regulatory Stakeholders*

As shown in Chap. 1, the transport sector shows a trend of increasing volumes and rising CO_2 emissions. The associated external effects represent an unresolved, but urgent issue for the planet, society and the economy. According to a survey commissioned by the international union of railways (UIC) from INFRAS and the Institute for Economic Policy Research (IWW) of the University of Karlsruhe, in the year 2000 road transport caused approximately 544 billion Euros in external costs in Europe, averaging 88 Euros per 1,000 tonne-kilometres. Governments and politicians have tried to tackle this economical, environmental, and societal issue for many years – with only partial success, hence the emergence of new stakeholders seeking to influence regulations – NGOs, plaintiffs, self-appointed watchdogs[5] (e.g. bloggers), local communities, industrial associations, and even industrial and commercial competitors.

[3]Esty and Winston (2009), p. 154.

[4]Thomlison (1992), Freeman et al. (2007), p. 7.

[5]Esty and Winston (2009), p. 290.

It is not within the scope of this chapter, or indeed this book, to analyse the many, many regulations covering logistics, transport, and supply chain management. Rather, I want to emphasise the importance for supply chain managers of understanding the trends and general characteristics of the regulatory environment affecting them. The regulatory environment of supply chains can classified by a morphologic box as shown in Fig. 5.3 below: Supply chain management decisions are made at several different levels (see Fig. 2.18 in Chap. 2). Hence, supply chain managers are increasingly concerned with a wide range of regulations – directly and indirectly.

At product level, relevant regulations may be product- and packaging-related. Although compliance with product-related regulations such as the EU directive "Restriction of Hazardous Substances (RoHS)" is primarily the responsibility of manufacturing departments, such regulations still present big challenges for supply chain managers; for example, end-to-end traceability, inventory control, potentially enormous excess volume and obsolescence costs, supply chain compliance at first to n-tier suppliers, and contractual obligations and liabilities related to legislative compliance. Third party logistics service providers face addition challenges in this regard. Typically they deal with the products of different industries, and therefore must tackle possibly different product-related regulations and also their interrelationships, such as those embodied in regulations prohibiting specific product combinations in transport and storage.

At supply chain network level, the main regulatory issues concerning supply chain managers in larger firms or exporters are raised by the diversity of regulations in different countries. The geographic range of governmental regulation extends from local planning commissions via regional, state, and national governments to global, supranational and intergovernmental organizations such as the United Nations or European Union. A recent example of an international, intergovernmental agreement of relevance to supply chains is the Kyoto Protocol which entered into force in 2005, establishing binding requirements for the reduction of certain greenhouse gas emissions (GHG) at national, and consequently, if indirectly, at company level.

Supply chain management and logistics are cross-company and cross-industry functions, hence the regulations and standards affecting them at process,

Complexity dimension							
Supply Chain Decision Levels/Elements	Product	Process	Technology	Infrastructure	People	...	
Geographical Scope of regulations	Local	Regional	State	National	Supranational	...	
Regulatory Instruments	Laws	Directives	Rules	Bans	Requirements	Ecotaxes	...
Regulatory Stakeholders (directly, indirectly)	Governments	NGOs	IGOs	Plaintiffs	Competitors	...	

Fig. 5.3 Regulatory environment of supply chains – morphologic box

managerial and technical levels are numerous and diverse. Examples concerning sustainability issues for supply chains are e.g. the carriage and handling of goods, employment and working conditions, road safety, drivers' hours, health and safety, pollution prevention, environmental protection, noise protection, and more besides.

Besides these complexities there is also a growing variety of regulatory instruments directly or indirectly related to logistics and transport: laws, directives, technical specifications, bans, rules, ecotaxes, and – a recent addition – emissions trading in the EU aviation sector.

So far, we can conclude that the regulatory environment of supply chains has become highly complex and increasingly challenging for supply chain managers. The associated stakeholder management issues show similar patterns of growing complexity. In line with the increasing range of multi-layered regulatory issues, the variety of stakeholders involved and their ability to communicate and exert influence are also on the increase. We discuss some of their characteristics below.

Non-Governmental Organisations (NGOs). Worldwide there are many thousand national and international NGOs, mainly dealing with social or environmental issues. Many large NGOs have established themselves as multinational organisations with wide reach and influence. Well known examples include Greenpeace, the Environmental Defense Fund, the Natural Resources Defense Council, to name but three.

NGOs follow many different paths to achieve their missions. These can vary between fully confrontational and cooperative, partnership-based routes. Along the confrontational route, for example, NGOs may call for new regulations or standards, and supported by media, educate the public to demand environmentally- and socially-friendly solutions. Adopting a more collaborative stance, they may support industry by becoming sources of best practice that may not be well known beyond a few pioneering companies. Or, as in the case of FedEx and the Environmental Defense Fund described below, NGOs can even enter partnerships aiming to collaboratively develop new technologies, for example. In 2000, FedEx Express and the Environmental Defense Fund joined forces to develop a diesel-electric hybrid delivery truck that dramatically cut emissions while increasing fuel economy. This was an ideal application for hybrid technology, given the frequent stops and starts of FedEx's 30,000-strong parcel delivery truck fleet, and the technology subsequently attracted further investment elsewhere. The FedEx-Environmental Defense Fund shared common objectives, and the latter was the catalyst in developing a competitive process for manufacturers to develop the technology.

Communities and individuals. Logistics and transport have multiple social and environmental impacts on local communities. These may involve ongoing operations, the sites of new logistics depots, or the expansion of existing facilities. Road transport, inherently an issue for many local communities due to the noise and pollution it creates, may become more important when large trucks are deployed in a 24-h operation. Similarly, freight airports operating 24 h clearly have negative local impacts. The siting or enlargement of ports, warehouses and distribution centres can be major issues for local communities, due to noise, congestion, intrusion on the landscape, and the use of scarce land.

Of course logistics locations can also benefit communities by creating jobs and opportunities for local business. But some community groups argue that there is a clear imbalance between the benefits and the costs to the local community of having larger and larger logistics locations, especially if those turn more into transition hubs with highly automated systems.[6]

Meanwhile, local community groups, and individuals at large within and beyond regional or national boundaries, are being aided by technology in making their views known. Web 2.0 internet development is creating greater connectivity and stronger, more varied communications platforms for communities and individuals. Channels for expressing views and exchanging information, e.g. with NGOs, have already multiplied: from blogs to polls to forums, videos (Youtube), social networks (Facebook) and attention-seeking (MySpace). "By adding interactivity to information, Web 2.0 empowers the individual and multiplies his or her collective power."[7]

Also NGOs make increasing use of Web 2.0, and integrate interest groups by providing online community technologies to support protest campaigns, which individuals can also start and manage online themselves.

Consumers. Consumers are in a much stronger situation than in the past, and demand more and more information about product and service content, and information on use, disposal, recycleability, and environmental impacts, such as embodied CO_2 in products, CO_2 emissions for vehicles, or services such as airline flights. The tools for calculating CO_2 emissions are still in their infancy, and the trend of emissions disclosure is likely to increase as the tools for measuring them increase in maturity.

Consumers are increasingly able to compare products and services, supported by NGOs, consumer protection authorities and commercial online comparison facilities. They are increasingly demanding products, packaging and services that are competitive in being friendly to the environment and to society; this in particularly evident in specific sectors and products such as food, toys, clothes, etc. Some consumers are also willing to pay higher prices for environmentally and socially sound products.

Companies and market researchers have already recognized a new consumer segment that they have called "LOHAS" (Lifestyles of Health and Sustainability). LOHAS consumers are interested in products and services that are healthy, safe and fashionable, but which also boast low environmental and negative social lifecycle impacts. LOHAS consumers, sometimes called "Lohasians", are usually above-average earners, and are found in several market segments.

Competitors. In the context of supply chain sustainability competitors represent a two-edged sword. On the one hand they may be willing to cooperate in certain logistics processes – transport and distribution, for instance – to create win–win situations; on the other hand, competitors can become environmental technology or process leaders who may set industry norms or establish legal mandates. Currently the latter case has been demonstrated in the context of supply chain carbon footprint

---

[6]Waters (2007), p. 245.

[7]Business and Poverty (2008), p. 176.

measurement. Many large 3PLs have started initiatives to develop, or least support, this development process. But even the global players find it hard to go all the way alone. The increasing complexity of the business environment and the increasingly rapid pace of change give priority to managing multi-stakeholder processes. Staying with the carbon emissions issue, the Greenhouse Gas Protocol (GHG Protocol)[8], one of the most widely used international tools used by government and business leaders to quantify and manage greenhouse gas emissions, demonstrates this point. A broad, multi-stakeholder process involving participants from businesses, policy-makers, NGOs, academics and other experts and stakeholders from around the world is dedicated to developing a new standard for supply chain GHG emissions measurement and reporting.

5.2.2 *Primary Stakeholders*

Managing the traditional players in the supply chain field – customers, suppliers, logistics service providers, and operators – lies at the core of all successful supply chain management strategies. That has not changed, regardless of the growing importance of sustainability issues. As we have seen in the previous sections, large, high profile companies often come under substantial pressure from a wide range of stakeholders to address environmental and social concerns. The main causes of concern are the companies' suppliers' and customers' activities, because with the extended producer responsibility featuring in today's business environment, companies cannot shift environmental or social responsibility problems up or down their supply chains, from the furthest upstream supplier to the most remote downstream customer. What happens outside your organisation itself may be what counts most for your secondary stakeholders. Hence, this reality requires that you ask new kinds of question: Are workers at your suppliers' factories exposed to toxic substances, or other unhealthy conditions? Do your suppliers dump hazardous waste in the local river? Do your customers discard your product in a way that causes litter or pollution?[9] Was the truck that had an accident and killed a mother and two children the truck that was shipping your goods?

The bestLog research confirms that given this pressure on high-profile companies, many of their suppliers lack incentives to seek innovative solutions. Hence the mostly non-regulatory pressures on suppliers, rather than regulatory influences, have proved to be the key drivers of environmental and social supply chain innovation. Because inter-company environmental policies are generally not

---

[8]An initiative by the partnership between the World Resources Institute (WRI) and the World Business Council for Sustainable Development (WBCSD).

[9]Esty and Winston (2009), p. 154.

addressed directly by regulations, companies are legally responsible for their own activities and not necessarily for their suppliers' or customers'.[10,11]

Having seen the external supply chain players and knowing what impact they may have on your supply chain, we will now cover briefly a number of action fields to manage supply chain stakeholders.

5.3 Action Fields for Supply Chain Stakeholder Management

The bestLog best practice case study analyses and interviews with practitioners, industry experts, industry association representatives, academics, and politicians revealed the following action fields for supply chain stakeholder management. We will take a brief look at these here before discussing them in more depth in the context of customer and supplier relations in Sect. 5.5.

Think Locally, Act Globally Executives in supply chain best practice companies often develop strong personal relationships with their stakeholders; thus they gain a very thorough understanding of how stakeholders in different countries and different sectors perceive their company's sustainability performance, and of their expectations. This allows these best practice companies to use communication and collaboration strategies customised to suit individual supply chain partners.

Combine Stakeholder Communication with Expertise and Innovation Best practice companies do not outsource stakeholder relations to lawyers or consultants. Stakeholder relations at best practice companies are run by experts and colleagues from the relevant business units, who fully understand the operations involved and focus on innovation and continuous improvement.

Communicate and Involve Communication and involvement is a broad action field, which entails reporting, providing information and educating, and forestalling confrontational behaviour; it also includes efforts to understand the needs, preferences and concerns of local communities and NGOs. As mentioned earlier, Web 2.0 internet development is creating greater connectivity and stronger, more varied communications platforms for communities and individuals. Best practice companies make use of these and other channels for communication and stakeholder involvement – for example, focus groups, opinion polls, formal progress review meetings, multi-stakeholder networks, websites with open forums, newsletters, canvassing of local businesses, community information displays, etc.[12]

---

[10]Hall (2006), p. 236.

[11]Porter and van der Linde (2008).

[12]Epstein (2008), p. 178.

Collaborate and Cooperate The involvement of stakeholders should be reciprocal. Best practice companies propose collaboration and partnerships based on enduring commitments. These may range from partnerships promoting environmental and social education, legislation, environmental and social programs, to employee voluntarism, and to working groups. Working groups can be particularly valuable, since their potential generally extends beyond information-sharing; they can identify and develop common strategies to moderate impacts and foster new opportunities for certain supply chain activities. Working groups can be developed so as to undertake significant joint projects or to form long-term partnerships.

Best Practice companies take often a proactive role and establish new types of relationships with regulatory stakeholders, even with competitors, where industry-wide solutions make sense (see the Cargo Domizil case). They aim to influence regulations and standards at the earliest opportunity where they affect their supply chain strategies.

Case:
Cargo Domizil – Coopetion in Intermodal Part-Load Transport
Three competing road haulage contractors in Switzerland were impacted by the government's modal shift policy; specifically, by a national night trucking ban (22:00–05:00 h) and by relatively high per-kilometre tolls for trucks on all Swiss roads.

The three competitors decided to turn a formerly state-owned loss-making business into a profitable private company, delivering combined mode services competitive with road haulage services, and capitalising on the advantage of overnight rail movements.

(See Case Collection in Part IV of this book)

Force a Value-Added, Strategic and Holistic Approach A common belief is still that there is an inbuilt, immutable trade-off between regulation and innovation. But there are best practice companies that have anticipated, for example, pollution, and used methods such as material substitution and closed-loop supply chains to eliminate or reduce pollution before it occurs.

What do these best practice firms do differently? The bestLog research reveals clearly that they consider the economic and social concerns of their stakeholders strategically and accord top management commitment to them. They apply the concept of value-added at both financial and operational levels, which leads to a totally different sustainable supply chain philosophy. This approach was embodied in Total Quality Management (TQM) in the 1980s and 1990s, especially in the automotive sector. Best practice companies consider end-to-end supply chains as value chains, and think about unexplored resource inefficiencies and opportunity costs at the level of suppliers, channels, and customers. At customer level, for example, resource inefficiencies show up in the use of the product, in discarded packaging, and in resources left in the used-up product.[13] Seeing supply chains as value chains allows best practice companies to see regulatory stakeholders as innovation drivers; they are motivated to collaborate with regulators to shape incentives and develop sustainable supply chain solutions (see the Wincanton case).

Case:
Wincanton Plc – Return Logistics – Implementation of WEEE Directive
The Waste Electrical and Electronic Equipment WEEE directive (Directive 2002/96/EC) was accepted in 2002, while in the UK was introduced in January 2007. The directives required that in all European Union countries, electronic and electric equipment will be collected, then reused or recycled. Wincanton's strategic approach and top management commitment resulted in the first complex WEEE solution offered on the UK market. Wincanton's value-added, strategic, and holistic approach comprised of an one stop solution that links logistics infrastructure and competence with electronic goods treatment services, a new model for return logistics in partnership between Wincanton and electronic goods retailer Comet, and new types of WEEE containers for electronic equipment collection allowing secure and monitored way of computers recycling.
(Piotrowicz (2008) Collection in Part IV of this book)

Eliminate Trade-Offs Some best practice companies go a step further. Rather than just seeking ways to mitigate the trade-off or conflict between regulation and supply chain performance, they try to eliminate it entirely. How significant an advance this may represent, even at operational level, is shown by the Daimler AG case study.

The bestLog research also observed that best practice companies were able to reduce their time-to-market because they did not need to submit their new product or manufacturing solutions to regulatory approval processes.

Case:
Daimler AG – Adaptive Logistics
Daimler realized that a new safety regulation associated with impaired supply chain performance was creating a niche market. This regulation concerning the means of securing loads drove Daimler to develop an innovative load security standard. Erich de Vries, the head of load security of Daimler, was surprised: When we began our review with the aim of defining a standardised

(continued)

[13]Porter and van der Linde (2008), p. 132.

procedure for securing loads, we couldn't have imagined that the outcome would generate such a positive response and such widespread interest in the transport logistics field.

(See Case Collection in Part IV of this book).

Track and Monitor Best practice companies track regulatory developments, monitor government policies, and competitors' initiatives, because they do not believe that all opportunities for innovation in their supply chains have already been identified. They recognize that regulatory stakeholders represent an effective source of new ideas.

The bestLog research identified further action fields, mainly related to customer and supplier relations, which we would like to discuss in greater in Sect. 5.5. As mentioned earlier, it is non-regulatory pressures rather than regulations that have proved to be the key drivers of sustainable supply chain innovations.

5.4 Determining the "Right" Supply Chain Stakeholders

In recent years, many companies have dramatically increased the quality and quantity of their regular interactions with stakeholder groups. Some have established "community panels" in order to learn about public concerns, but, too often, companies have been taken by surprise by sustainability issues, and have suffered organizational crises and related costs accordingly.[14] Hence, determining who are the most relevant and important supply chain stakeholders is significant for your strategy development, as shown in Chap. 2. Understanding the interests and the potential impacts of each group on the supply chain is invaluable. This section describes an approach (see Fig. 5.4 below) to defining who the most relevant stakeholders in your supply chain are.

Step 1. Distinguish primary and secondary SC stakeholders – apple or orange?

The first step in this process is to distinguish between primary and secondary stakeholders, as described earlier because the framework needed to assess your dependence on primary supply chain stakeholders is different from that regarding secondary ones, as described below. But the importance of secondary supply chain stakeholders should not be underestimated, since they may have a significant indirect effect on the supply chain through their interactions with primary stakeholders.

[14] Epstein (2008), p. 178.

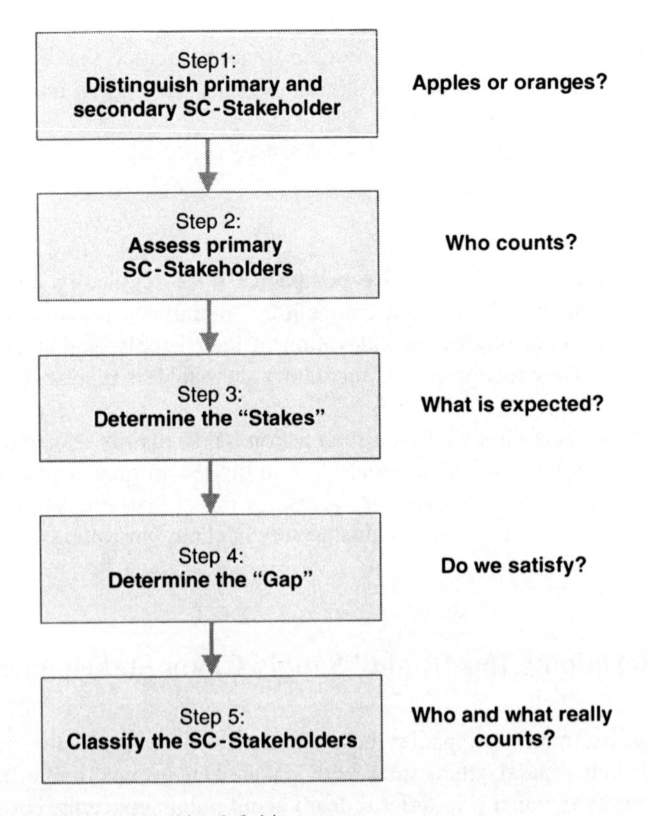

Fig. 5.4 Supply chain stakeholder definition process

Step 2. Assess primary SC-stakeholder – who counts?

The following framework is designed to help assess your primary supply chain stakeholders. It is based on the assumption that the following interlinked attributes of primary stakeholders determine their importance, and in consequence your dependence on them (Fig. 5.5):

- Their access to supply chain knowledge and information
- Their substitutability (replaceability)
- Their access to networks and infrastructure
- Their value creation along the supply chain

Access to supply chain knowledge. Access to Supply Chain Knowledge is nowadays a very important assessment criterion. The basic supply chain management theory is about information sharing along the whole supply chain. This is the fundamental axiom of the supply chain management concept. One key knowledge area is the knowledge about customer requirements. A good example is the fashion retail chain Zara. Zara is able to create every year around 10–12 fashion collections; more than the average in the industry. Their product development is strongly

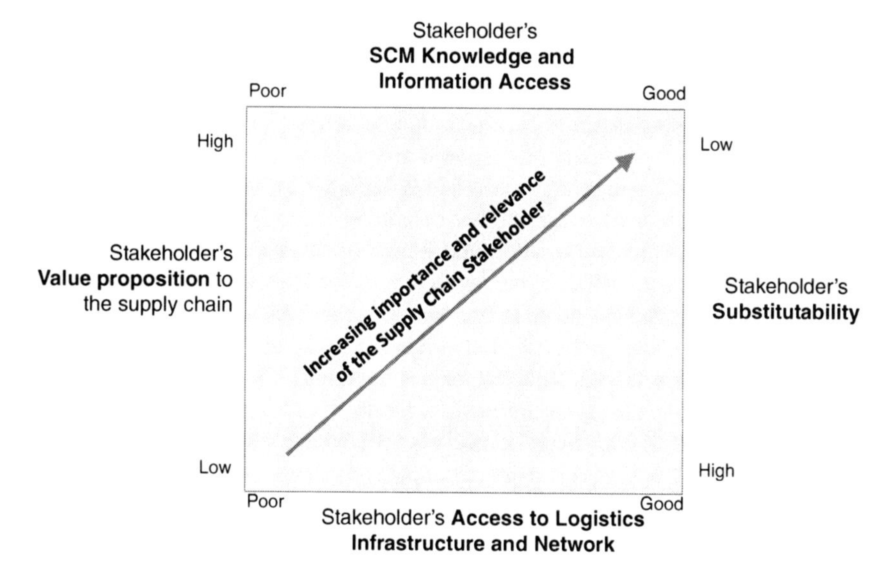

Fig. 5.5 Framework to assess primary supply chain stakeholders[15]

customer driven and their supply chain is very responsive in terms of the time-to-market. One reason is, compared to many competitors, the strong vertical integration in the supply chain. But the key, in our context, is their structured collection of customer requirements in the retail shops, which are quickly (on a daily basis) communicated to fashion designers and all other relevant supply chain partners along the end-to-end supply chain.

Another good example is a failing transport outsourcing project at an SME industrial laundry Haake GmbH in Berlin. This company has long relationships with many of its customers, e.g. hotels and foster homes. The process of delivering clean laundry and the collection of dirty laundry was outsourced to a renowned transport operator in the Berlin region. Soon after the outsourcing Haake received plenty of customer claims. The main issues were the wrong collection, wrong washing procedures, and wrong storage at the customer locations on-site. The main reason was simply the continuous change of the drivers. The collection and delivery of the laundries requires the on-site knowledge of customer specific processes and structures in order to find and sort the right laundries with the right washing requirements. Since, the customers do not want to organise the on-site processes and do not organise any logistics operations. The transport service provider could not manage this knowledge as they their changing route planning lead to different drivers processing the customers each time. Haake stopped the outsourcing project and in-sourced the process back.

[15]The four axes in the figure are not connected. For example: good access to knowledge does not necessarily correlate with good access to infrastructure. The figure just summarises the relations of the four axes with the importance of a supply chain stakeholder.

The more knowledge your stakeholders own about your value chain the higher their relevance with regard to your supply chain stakeholder management.

Substitutability. Substitutability considers whether you are able to change a certain primary supply chain stakeholder by another one, who can basically offer the same services or products; in the same quality and to the same price. This is particularly relevant for regular, standard load haulages by carriers, who can simply substitute on the spot market. Or another example are 3rd-tier suppliers for C-parts as for example standard screws and nuts. Where in contrast, first-tier suppliers for A-Parts are much more difficult to change due to the knowledge, capabilities, and resources they own and the value they create in your value chain.

The lower the substitutability of a primary supply chain stakeholder is, the higher its importance in your stakeholder management.

Access to infrastructure and network. Transport and logistics efficiency basis primarily on Economies of Scale, especially in lean and physical-efficient supply chains. In transport operations, as an example, this depends often on the network you serve or the network you own. The basic rule is: The larger the transport networks and volumes the higher the bundling potentials and in consequence the higher the utilisation per transport and the lower the costs per transported unit. Many transport service providers achieve competitive advantage because they have access or own such networks and in consequence are able to offer shippers better prices by lower costs.

Similar case with infrastructure: Certain stakeholders like port operators or railway operators are in a unique position to be a quasi monopole in terms of infrastructure access. Although transport and logistics markets are liberalised in many countries worldwide, certain transport and logistics markets do not work by traditional market mechanisms. Hence, certain stakeholders own these positions and own the access to infrastructures which may be highly relevant for your supply chain operations.

The higher these accesses of your stakeholders, the more important they are with regard to your supply chain stakeholder management.

Value contribution. A supply chain represents actually a value chain, where the value of a product or a service, measured separately by price and by costs, increases along the supply chain (see following Fig. 5.6).

This example of a value chain shows that primary supply chain stakeholders create different values (in total) in the supply chain, depending on the materials they produce and deliver and the logistics services they fulfil. Of course, the term "value" and its measurement can be debated, and price and cost are clearly factors. Consider the following: costs and value-added are very different within different processes. For example, the total costs (materials and process costs) involved in assembling a Porsche badge on an almost-finished Porsche 911 Turbo are much lower than the price that this assembly process justifies. Or would you pay 150,000 Euros for a Porsche with no Porsche badge?

Value chains give you a further criterion for assessing a stakeholder's importance; the greater its value contribution, the greater will be its importance in your supply chain.

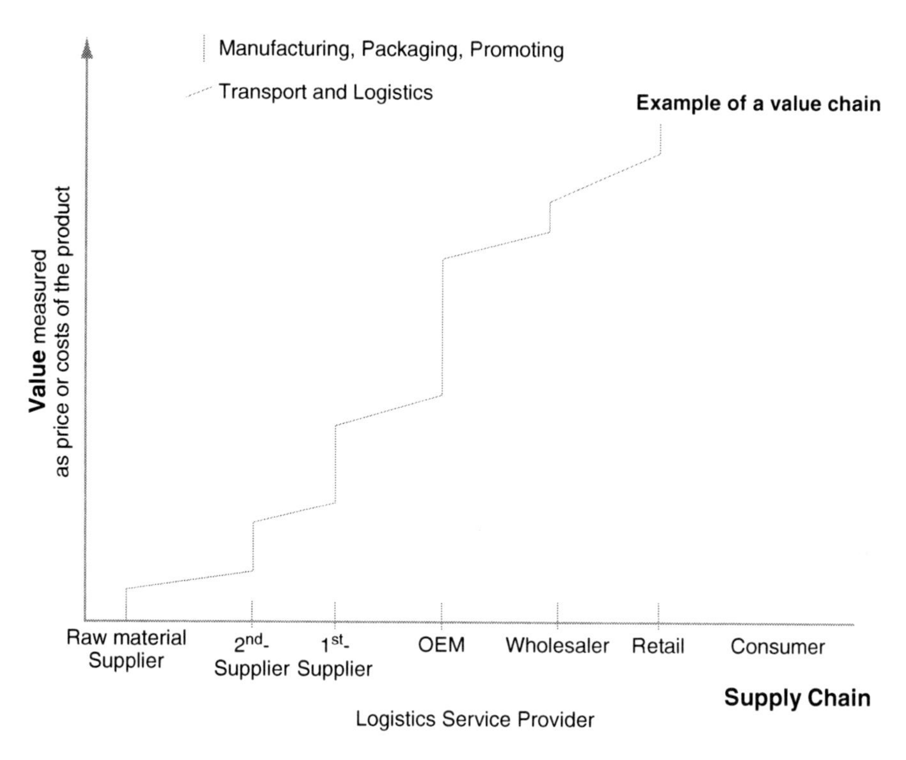

Fig. 5.6 Example of a value chain in a producer company's supply chain

This framework for brief, simplified supply chain stakeholder analysis will enable you to list primary supply chain stakeholders in order of priority, which leads us to the next step in understanding their specific demands, or desired returns on their "stakes".

Step 3. Determine stakeholders' "stakes" – what do they expect?

Supply chain managers must determine each stakeholder's "stake" in relation to whatever issues may be under consideration. Due to the great complexity of stakeholders' multiple[16]-relationships within a supply chain, this requires ongoing structured monitoring, forecasting,[17] assessment, and even proactive efforts towards two-way communications to achieve accurate understanding of the stakes involved. BestLog research reveals that a healthy sensitivity to primary "stakes" and appropriate communications with stakeholders (as required[18]) are essential

[16]One stakeholder may have different stakes in some issues; according to Polonsky (1995), p. 35.

[17]Remember the "scenario planning" concept in Chap. 2.

[18]Too much communication with certain stakeholder groups may be counterproductive, given the risk of issues ultimately gaining public attention.

success factors; we will return to supply chain stakeholder management strategies later on.

Sometimes the "stakes" of several groups of stakeholders may be more important than that of a single stakeholder. For example suppliers sharing the same interests may form a "coalition" to increase their influence or power, or interest groups may cooperate with media to enforce their direct and indirect pressure. It is important to understand the structure of your supply chain stakeholder network, and how its members communicate.

Understanding stakeholder's expectations or demands in a dynamic business environment isn't easy, particularly in long, international supply chains. There are two main reasons why:

(a) Expectations may not be clearly defined or reflected in actions in a way to which companies can react effectively. For example, the measurement of supply chain carbon footprints in relation to the CO_2 emissions market planned by the European Union has to date not been specified or standardised.
(b) Communication between your firm and your stakeholders may not be conducted in the desired form, or it may exist, but is used by the parties to threaten each other rather than as an aid to cooperation.

BestLog interviews with practitioners show clearly that personal contacts based on honest, structured communication with realistic commitments are key success factors in understanding stakeholders' interests, and in the end, in establishing more cooperative, less hostile relations with primary and secondary supply chain stakeholders.

Step 4. Define the "gap" – are we satisfying stakeholders' needs?

Clearly understanding stakeholder requirements, combined with understanding of the strengths and weaknesses identified in your strategy development approach,[19] allows you to measure the "gap", so as to meet the stakeholder's requirements.

The clear evaluation of this "gap" depends mainly on overcoming the aforementioned lack of clearly specified and quantified "stakes". Imprecise and purely subjective understanding of stakeholders' needs can lead to a high risk of adverse behaviour, magnified by unfair media reports, for example. So, the greater your lack of insight into stakeholders' priorities, the less effective will be your ability to meet their expectations efficiently.

Step 5. Classify the SC-stakeholders – who, and what, really counts?

Step 5 aims to classify your supply chain stakeholders to provide the basis for a focused, differentiated action plan for sustainable stakeholder management. The classification results according to Savage from two important criteria:

(a) The stakeholder's potential to threaten your organisation and your supply chain
(b) The stakeholder's potential to cooperate with you

---

[19]See, Chap. 2.

The potential *threat* posed by a stakeholder is like the development of a "worst case" scenario. Of course, your dependence on a given stakeholder, which you have assessed in Step 2 for your primary supply chain stakeholders, determines that stakeholder's potential to threaten your company or the rest of your supply chain. But in this second step, the threat depends largely on the willingness of a stakeholder to cooperate, and on the opportunities a stakeholder enjoys. Put simply, dependence does not lead necessarily to threats; rather. It defines the *potential* threat. Hence you should, besides assessing dependency, also consider the capacity, opportunity, and willingness of a stakeholder to threaten your organisation or supply chain.[20] For example: if your organisation is seeking to improve delivery reliability and to reduce inventory by entering into a vendor-managed-inventory relationship with a supplier, then the supplier's opportunity to threaten your organisation and the following supply chain will be quite considerable.

Managers can assess the willingness of a stakeholder to threaten their organisation by considering the quality and durability of the stakeholder-organisation relationship.[21]

The potential for *cooperation* is the next chosen attribute for classifying stakeholders. Considering a stakeholder's potential for cooperation is like the development of a "best case" scenario.[22] This attribute fits very well to the supply chain management principle: communication, cooperation, and collaboration along the end-to-end supply chain.

The better you understand stakeholder's potential for cooperation, the more focused you can perhaps rethink and effectively adapt existing cooperation and collaboration models in your supply chain.

Chapter 2 mentioned the strategic SSC-programs "Innovation", "Progression", and in part "Re-Design", which required a collaborative approach going beyond merely defensive strategies.

Again, the stakeholder's capacity, willingness, and opportunities to expand its inter-dependence with your own organisation will determine the scope of cooperation, as well as the risks involved. But bear in mind that the more dependent the stakeholder on your organisation, the higher its willingness to cooperate is likely to be.

Taking both criteria[23] into account you can classify your organisational supply chain stakeholders into four types, as shown on the following Fig. 5.7: This typology helps supply chain managers to specify their sustainable supply chain

[20]Savage et al. (1991), p. 63.

[21]Savage et al. (1991), p. 63.

[22]Savage et al. (1991), p. 63.

[23]Two further stakeholder attributes allow you to refine your stakeholder classification in this regard: Legitimacy: "A generalised perception or assumption that the actions of an entity are desirable, proper, or appropriate within some socially constructed system of norms, values, beliefs, definitions" (Suchman 1995); and Urgency: "The degree to which stakeholder claims call for immediate attention." (Mitchell 1997).

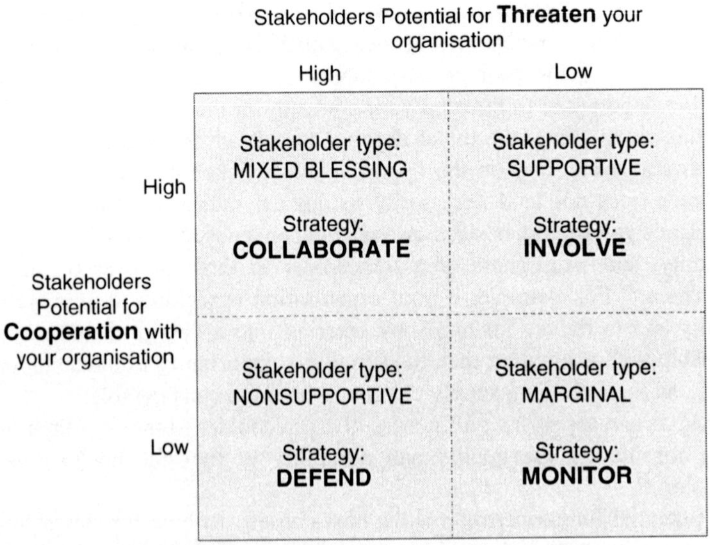

Stakeholders Potential for **Threaten** your organisation

Stakeholders Potential for **Cooperation** with your organisation	High	Low
High	Stakeholder type: MIXED BLESSING Strategy: **COLLABORATE**	Stakeholder type: SUPPORTIVE Strategy: **INVOLVE**
Low	Stakeholder type: NONSUPPORTIVE Strategy: **DEFEND**	Stakeholder type: MARGINAL Strategy: **MONITOR**

Fig. 5.7 Typology of organisational stakeholders (Savage et al. 1991, p. 65)

strategy[24] with generic strategies for managing supply chain stakeholders, a topic to which we will return in the next chapter.

5.5 Generic Stakeholder Management Strategies

The aim of this section is to desribe a framework to help managers translate their sustainable supply chain strategy into action with regard to stakeholders. The overall goal is to close the previously mentioned gap identified in step 4 of the aforementioned stakeholder definition process (see Fig. 5.4).

Savage's typology shown in Fig. 5.7 above defines four types of stakeholders and four associated generic strategies:

1. Stakeholder type: Supportive; Strategy: Involve
2. Stakeholder type: Marginal; Strategy: Monitor
3. Stakeholder type: Mixed Blessing; Strategy: Collaborate
4. Stakeholder type: Nonsupportive; Strategy: Defend

Here we will focus mainly on "Collaborate" and "Defend" strategies, since the attitudes and actions of "mixed blessing" and non-supportive stakeholders tend to be the most difficult to respond to effectively, and they may have direct economic impacts on the supply chain, and pose risks for the company's public reputation.

[24]See Chap. 2.

5.5.1 *The Mixed Blessing Stakeholder and the Collaborate Strategy*

Mixed blessing stakeholders play a major role in supply chains; their potential to threaten the supply chain and their potential to cooperate are both considerable. Typical stakeholders of the mixed blessing type could be customers or clients, first- to n-tier suppliers (including service providers), and other companies and competitors with complementary products and supply chain systems[25] Product characteristics determine fundamentally different supply chain strategies according to Fisher, see also Chap. 2.

Bear in mind that these stakeholder categorisations may be debatable. The mixed blessing type is generic, and individual stakeholders must be assessed in the real-life context. Not every stakeholder in a given stakeholder group will necessarily be willing to cooperate with other supply chain members. For example, competitors in certain sectors are likely to be unwilling to consider cooperation in specific fields such as R&D, although in some contexts the degree of horizontal cooperation between competitors, even in R&D, should not be underestimated. There are for example many collaborative R&D efforts in high-tech sectors. Logistics and supply chain management are in many areas particularly appropiate areas for "coopetion" (cooperation between competitors), because they seldom involve much risk in relation to the disclosure of competition-critical information.

The same goes for customers and first-tier suppliers. If your company's services can easily be exchanged for another's by a given customer, you cannot expect much enthusiams for cooperation from that customer. Some first-tier suppliers, like Bosch in the automotive sector, for example, have enough power in certain product areas to refuse to join in cooperative activities unless they offer a win–win situation for them. Mixed blessing stakeholders may become either more or less supportive.

The mixed blessing stakeholder may best be managed through collaboration, as stakeholders involved closely in cooperating and collaborating with your company will find it more difficult to act against your organisation's interests. Collaboration increases your stakeholders' dependence, and reduces their willingness to threaten your interests.

5.5.2 *The Non-supportive Stakeholder and the Defend Strategy*

Non-supportive stakeholders, who score high as potential sources of potential threats but low on potential cooperation, are particularly challenging for supply chain managers. Typical stakeholders of the non-supportive type could include

---

[25]Regarding general typology see also Savage et al. (1991), p. 67.

competing companies, media, trade unions, governments[26] and international regulators.

Again, we should emphasise that assignments of individual firms and organisations to this non-supportive category may be open to debate. The non-supportive type is generic, and individual stakeholders must be assessed in the real-life context. Since, for example, policy makers may rank high as potential sources of cooperation with regard to certain issues, such as, for example, carbon emissions measurement. They may be willing to understand the needs of practitioners, to collaborate, and to invest in dedicated R&D projects. The bestLog project is a good example of just such collaboration, as it fosters learning and promotes the exchange and dissemination of knowledge and between business interests and policy makers right across Europe. Non-supportive stakeholders, then, may in certain circumstances become more supportive.

The non-supportive stakeholder may best be managed via a defensive strategy, one that aims to reduce dependence on the stakeholder concerned. We will not discuss this strategy in depth for the moment, as it will be covered indirectly in Chap. 6. The following section will now discuss further the "Collaborate" strategy for sustainability, in particular with "customer" and "supplier" stakeholders in mind.

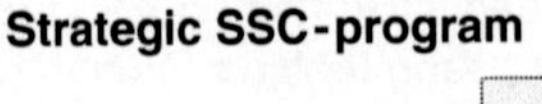

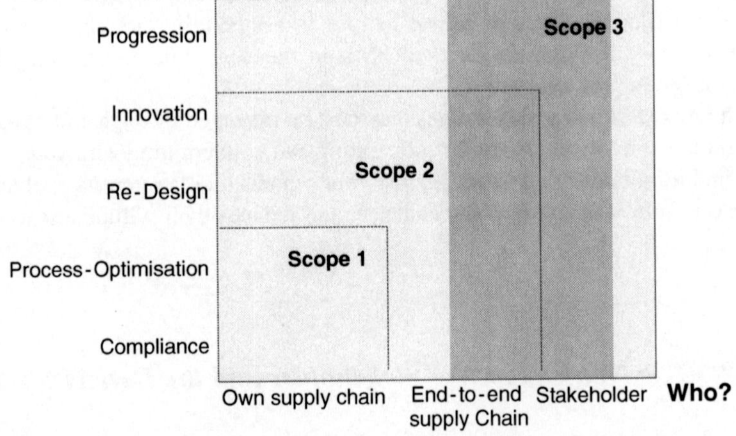

Fig. 5.8 Action Fields to manage primary supply chain stakeholders

[26]See also Savage et al. (1991), p. 66.

5.6 Action Fields for Sustainability in Customer and Supplier Relations

In Chap. 2 we saw how supply chain managers can extend or even re-design their existing supply chain strategy in pursuit of sustainability. How radical this approach is, which is specified by the implementation level and the Strategic Sustainable Supply Chain (SSC) Program (see Fig. 5.8 below), determines the action fields for the management of supply chain stakeholders, shown by the light blue bar. The bar extends over scope 2 and 3, because suppliers and customers may be far upstream and downstream respectively from your company. The more closely a supplier or customer is integrated with your supply chain and the more you can dictate to your supplier and customer, the more closely it corresponds to scope 2, and vice versa. This continuum relates again to the mixed blessing stakeholder, who could become either more or less supportive in terms of collaboration.

Taking this into account we can create a matrix combining the Strategic SSC-Program and the generic Stakeholder Management Strategies (Fig. 5.9). Again, this matrix is generic, and serves as a template to be customised to suit the specific conditions of your sector, your company's size and geographic scope, and the supply chain issues under consideration.

The following pages describe various action fields and case examples to provide an overview of some theoretical approaches as well as some practical insights from the bestLog research with regard to sustainable SC-stakeholder management. The focus will be on the "Collaborate-Strategy", with reference to both customers and suppliers.

5.6.1 Customer-Relationship

"Customers provide the lifeblood of the firm in the form of revenue."[27]

This quotation remains as valid as ever in today's highly competitive, global business environment. Relationships with your customers deserve the closest attention in your company.

Action fields for sustainability in customer relations are multifaceted. The exploration of appropriate concrete actions depends on your level of integration within the supply chain. The more closely you are involved in the long-term decisions of your customer's supply chain, in general the greater will be your leverage, and the potential for sustainable solutions.

[27]Freeman (1994), p. 43.

Strategic Sustainable Supply Chain Programs

Stakeholder Management Strategies

	Compliance	Efficiency	Re-Design	Innovation	Progression
Collaborate					
Customers	• Fulfil cust. standards • Create transparency about your standards and processes • Recovery planning for customer-side processes	• Closed-loop supply chains (processes) • Synchronise inbound- and outbound flows • Optimise, standardise, and integrate processes	• Customised decoupling of transport and warehousing • Closed-loop supply chains (structure)	• Develop new sustainable products and services • Design for Logistics • Carbon Footprint • Sustainability Labels	• Collaborative research and promotion
1st*-to n-tier suppliers	• Request, control legal and industry standards • Develop an own SC sustainability standard • Measure and assess sustainability performance	• Closed-loop supply chains (processes) • Synchronise inbound- and outbound flows • Optimise, standardise, and integrate processes	• Re-Design procurement structures • Regionalization • Assess location sustainability	• Collaborative product re-design and development • Evaluate new outsourcing potentials • Develop new incentive systems	• Develop best practice and knowledge platforms for your suppliers • Enable supplier cooperations
Complementary** Companies and Competitors*		• Integrate logistics and transport processes • Collaborative procurement	• Share infrastructure and service providers • Consolidate network structures	• Collaborative R&D	• Develop best practice and knowledge platforms for the entire sector • Develop new standards
Defend					
Governments	• Monitor regulations • Fulfil legal standards • Monitor own marketing and media activities	• Understand cause-effect relations between regulations and process performances	• Assess supply chain network re-design with regard to regulations	• Develop new and alternative technologies	
1st-tier suppliers-single sourcing		• Implement recovery processes	• Assess sourcing and supply concepts • Implement recovery structures		

Fig. 5.9 Action fields matrix to manage supply chain stakeholders and some examples

The bestLog research reveals the primary importance of three major action fields corresponding to the level of integration, in ascending order of integration:

1. Compliance and transparency in supply chain processes
2. Realising new efficiency potentials via process-optimisation and supply chain re-design
3. Product and service development and associated cost- and gain-sharing models

Below we will take a closer look at the related capabilities identified within best practice companies during the bestLog research.

Compliance and Transparency in Your Customer Relationship Your customers' supply chains may become increasingly complex and vulnerable, e.g. due to internationalisation or increased outsourcing efforts. Thus your customers, retailers for example, may want to control their supply chain risks, (for example, the breach of ethical norms by suppliers' employment of child labour), by requesting compliance with legal, industry, or internal standards.

This represents nothing new in today's supply chains. But bestLog clearly shows three novel challenges in this context: (1) The incorporation of totally new metrics such as CO_2 emissions into existing compliance reports, and the difficulty of obtaining credible data from next tier suppliers and contractors. (2) The need to find a balance between transparency and risk. Now, as ever, excessive transparency creates risk from competitors, regulators, and even customers, e.g. in terms of your cost transparency. (3) Understanding and supporting the compliance efforts of your customers, and finally, seeing customers' compliance requirements as opportunities for innovation.

The capabilities identified within best practice companies are analogous to these challenges:

1. Best practice companies understand the standards imposed by their customers and comply with them. Furthermore they control their standards implementation effectively and report regularly to their customers as required.
2. Best practice companies "do good" and talk about it! They improve their reputation as reliable partners by demonstrating compliance with all relevant legal and industry standards in customer-facing processes.

New Efficiency Potentials in your Customer Relationship Another situation, involving a higher level of integration: The complexity of your customer's supply chain is considerable, and your business is concerned with and affected by this complexity. For example: You are a supplier of A-parts and manage others of your customer's suppliers; you control a large range of supplies of mixed products, you control several interfaces with other 3rd parties, your coordination and communication efforts are considerable, and you are also dealing with international supply chains. A consequence of this complexity is that customer relations are likely to be associated with inefficiencies, rising costs, and wide environmental and social impacts.

Sustainability-related measures aiming to increase the efficiency of supply processes can be divided into two basic categories. First, there are the "quick wins", which generate significant results at low cost. These are usually cases of simple process optimisation, which do not involve changing logistics structures. They lie usually within the sphere of responsibility of logistics and supply chain managers and do not require cross-company approval. The second category of sustainability-related measures comprises fundamental process and structural changes, requiring greater investments – such as, for example, investments in alternative fuel technologies, new warehouses and/or warehouse equipment, implementation of environmental management (e.g. carbon footprint measurement) systems, or the implementation of cross-company fleet control systems.

Your efforts in exploring potential new efficiencies in customer-related processes will depend, of course, on the cost- and gain-sharing model you operate with your customer. If your customer is not willing to risk investment in fundamental changes, you should adopt a "Process Optimisation Strategy" rather than a "Re-Design Strategy".

Process Optimisation in Your Customer Relationship Planning, optimisation, standardisation, and integration of supply processes – especially IT-based processes – allow you to cut costs in the complex supply chain in order to remain competitive. But this is not new to supply chain managers; it's part of their day-to-day work, especially if they have adopted a lean supply chain strategy. The novel challenge in the context of sustainability is to explore new efficiency potentials in complex processes, driven by environmental goals which complement efficiency goals.

The bestLog research reveals that the highest potential for efficiency gains can be found at process interfaces with primary supply chain stakeholders along the product life-cycle. Here is one example which highlights best practice capabilities identified in this context: In the FMCG and retail sector, regulations governing disposal and recycling have resulted in high regulatory costs. Each firm in the supply chain is legally responsible for their own activities in this field, so the process interfaces between suppliers, customers, and disposal companies are driven essentially by regulatory pressures. Collaborative planning and process improvements in this field are uncommon, although best practice examples in the field of so-called closed-loop supply chains show high cost cutting potential.

Closed-loop supply chain management involves the reintegration of returned products into supply and manufacturing processes. It is based on holistic, far-sighted, collaborative planning of flows of goods in opposite directions, and on information shared within a network of suppliers, customers, and disposal companies. Closed-loop supply chain management extends the concept of reverse logistics, lead at operational level to synchronised supply and disposal flows, which in turn opens up new potentials for efficiency, for example, in bundling, volume smoothing, network utilisation, etc.

1. So, identify together with your customer and its other primary SC stakeholders further process fields in the end-to-end supply chain which may improve the

overall efficiency of the supply chain. Think beyond your own and your customer's processes.
2. As suggested in the strategy chapter: Looking beyond your own four walls for improvements requires that supply chain managers have adequate responsibilities and are offered effective incentives.
3. Identify the potential of state-of-the art ICT (Information and Communication Technology) to enable further efficiency improvements. Following the example above: Inbound and outbound transport flows are usually planned and executed separately, particularly if different transport service providers are in charge of each. Merging of both flows by synchronising them often fails due to an absence of planning data, and the dynamics of transport and route planning. An IT-driven system to exchange and synchronise transport data in real time is an essential success factor, together with a re-designed cost- and gain-sharing model.

Re-design in Your Customer Relationship There is a limit to the efficiency improvements that can be won if you do not change existing supply chain structures; see also Chap. 2, "Shift from defensive to aggressive strategic SSC-programs".

Here is an example which highlights best practice capabilities identified in this context: Creating inventory and additional warehousing is an example of structural change, and represents an approach which generally goes against the grain of current supply chain strategies. Today's supply chains are, as mentioned earlier, closely integrated and highly synchronised, especially for customers in the automotive or high-tech sectors. These supply chains are mainly geared towards low inventory levels and speed of operation. But do all your customers require speed? I am sure they do not, at least, not for all the products they procure. BestLog interviews with practitioners confirm this: not all customers require for all their products a 24- or 48-h delivery window. The reliability of supply is usually more important.

So why not rethink the potential of decelerating supply processes or of creating additional inventory volumes – for certain customers and products – in the logistics network, in order to slow down the pace of trade within the supply chain? The creation of these new stocks could enable you to separate transport operations from inventories, creating more leeway for consolidation and bundling measures in transport, and thereby improving efficiency (see the Tschibo, Hapag Lloyd case).

What bestLog research clearly revealed is that best practice companies overcome the key challenge of finding potential new efficiencies based on structural or process improvements: They have a very good understanding of different customer requirements and they are able to set up dedicated supply chains in order to balance customer requirements with their supply chain performances. Hence, rethink existing customer supply chain requirement and explore new win–win models together.

Case:
Tschibo – Hapag Lloyd. Speed Reduction of Container Vessels
A good example of customised speed reduction is provided by the ocean carrier Hapag Lloyd, which reduces the speed of certain container vessels and achieves a fuel usage reduction of 50% by reducing their speed from an average 25.0 knots to 20.0 knots, and a reduction of 30% through a speed reduction from an average 23.5 knots to 20.0 knots.

Global shippers such as the €3.2 billion German retailer Tschibo, which emitted 146,000 tonnes of CO_2 from 100,000 tonnes of sea freight in 2006, are already taking such savings into account, resulting in lead times extended by 3–5 days for shipments from e.g. Hong Kong to Hamburg, involving 26–28 days at sea.

Companies are starting to rethink their supply chain models, driven by increasing oil prices or by falling demand, for instance during the economic crisis that began in 2008. Long-term developments and trends, as in the above sea freight example, where container vessels became larger, leading to more capital invested per unit and consequently leading to higher average speeds, must be challenged by best practice companies seeking new efficiency potentials in their supply chains.

Sustainable Product and Service Development in Your Customer Relationship Suppose that your customers wanted to develop new, environmentally-friendly and socially responsible products and services and intended to step up their corresponding marketing activities.

Sustainability measures with regard to logistics primarily involve (re-)designing loading equipment, products, packaging, and logistics services aiming for higher utilisation rates and safer handling processes in transport and warehousing.

Interestingly, what the bestLog research shows is that the main reason why projects fail in this area are not shortcomings in technical capabilities for (re-)designing products, packages or services. Rather the underlying cost- and gain-sharing models in supplier–customer relations were inappropriate and betrayed a lack of incentives for suppliers.

Best practice companies take care to evaluate whether their customers are willing to invest or share in cost savings. And if not, they evaluate potential costs and benefits which they can exploit for their own operations, e.g. through improved transport utilisation.

1. Check the price sensitivity of your customers and consider possible differentiation strategies for new or re-designed products, packages, and supply chain services, especially if you are a logistics service provider.
2. Emphasise collaborative research if customers are willing to explore potential new efficiencies, and if they require highly innovative products and services

3. Emphasise the early involvement of the customer's logistics and supply chain managers in collaborative product and packaging development. This allows the early prevention of logistics inefficiencies due to unnecessary product or packaging properties, in what is known as Design for Logistics. The potential efficiency improvements in your logistics systems may allow you to offset the unwillingness of your customer to invest or share cost savings.

5.6.2 Supplier Relationship

"Suppliers are vital to the success of the firm…"[28]

As with customer relations, arenas for action to enhance sustainability in supplier relations are multifaceted. The exploration of concrete actions depends on the level of integration with your supplier. Have you considered a single-sourcing supplier? Is that your first-tier supplier for A-parts? Is it a commodity, a functional or an innovative product that you are sourcing? The more influence your supplier has in the long-term decisions affecting your supply chain, the greater should be your attention and willingness to cooperate (see also "Step 2: Assess primary SC-Stakeholder – Who counts?" mentioned in Sect. 5.4).

Analogous to the findings concerning customer relations, the bestLog research reveals three major action fields corresponding to the level of supply chain integration, in ascending order of integration:

1. Compliance and transparency in supply chain processes
2. New efficiency potentials from process optimisation and supply chain re-design
3. Product and service development and associated cost- and gain-sharing models

The second and third of these action fields are analogous to those described above with respect to customer relations. The key differences from the customer perspective mainly involve the related incentives. Several bestLog case studies have shown that sustainability actions identified and driven by suppliers are associated with clear, measurable win–win results. Hence best practice companies are nowadays often willing to rethink the business models they operate with their suppliers, particularly as concerns investment-, cost- and gain-sharing issues, which are the critical success factors for sustainable supplier collaboration. This is especially the case where action fields such as structural changes to supply networks, or product re-design, or new packaging, or fleet modernization are concerned.

Hence, next we will take a closer look at the "compliance and transparency" field of action and the related capabilities identified within best practice companies during the bestLog research.

[28]Freeman (1994), p. 43.

The situation in your supplier relations is basically the same as described in the above regarding customer relations. It simply involves a change of perspective. The difference is that you are concerned with the activities of your suppliers and may also be in a sense responsible for their products and services – for example, for their social behaviour in their manufacturing activities. Your aim is to ensure their compliance, at least with minimum standards and applicable laws. All this is highly relevant for you, if you are a large public company, listed on the stock exchange, and in particular, if you have direct contact with consumers, because such companies are usually more closely observed by social and environment interest groups than are, for example, B2B SMEs.

Measures to increase compliance and transparency in your supply chain (first- to n-tier) can be divided into two basic categories – reactive and pro-active. The reactive approach requires that all your suppliers comply with national and international legislation and with regulatory and industry-specific environment and social standards, e.g. with regard to health and safety, social conditions, child labour, waste disposal, etc.

The bestLog research reveals significant challenges in the implementation of compliance in supplier relations: (1) There is a lack of legal requirements and common standards covering specific issues for environmental and social sustainability in logistics and transport. (2) There are cultural differences within international supply chains which impede the smooth company-wide roll-out of social and environmental standards, especially in large corporations. (3) Great effort is required to verify implementation, particularly compliance with legal requirements, at supplier locations.

The capabilities identified within best practice companies include:

1. Best practice companies maintain a global perspective and track developments, understand the differences between different regulatory requirements and standards, and gain knowledge of each country's and each sector's specific issues.
2. Best practice companies try to develop incentives and to motivate their suppliers to invest in complying with standards, especially their first-tier suppliers (see also the Sharp case)

Case:
Sharp's Collaboration with the Green Transport Marketplace, SmartWay

Sharp Electronics Corporation is a global electronics manufacturing company based in Osaka, Japan. One of the company's core values/commitments is reducing environmental impacts including minimizing greenhouse gases. Sharp in the USA wanted to explore how they could contribute in the transportation area. Sharp decided to join a transport partnership – called SmartWay – which was implemented by the US governmental organisation EPA.

(continued)

Sharp must convince Sharp-approved carriers to join Sharp in becomimg an EPA SmartWay partner or risk having new carriers handle business, increasing risk.

The provided tools to the carriers to improve their processes led to commitments from carriers.

(See Case Collection in Part IV of this book)

3. Further, best practice companies motivate their first-tier suppliers to implement the same compliance culture and awareness, and to extend it to the second and subsequent tiers of the supply chain
4. And the most difficult, but most important: Best practice companies verify the results. They are able to prove whether or not their suppliers are applying the given standards and laws in practice, because the standards certification processes, particularly in respect of legislation, seldom check all relevant activities or check them continuously.

If you find that existing mandatory and voluntary standards do not close your compliance gap (identified in step 4 of the supply chain stakeholder definition process), take a proactive approach and develop and implement your own standards for your targeted supply chain sustainability, as large best practice companies do. These will be relevant to your own company and context, and to your own supply chain-specific strategy.

The bestLog project investigated this topic in depth, and analysed more than 100 standards, certificates, and awards in Europe. The aim of this investigation was to verify their potential applications for all three sustainability dimensions (economic, social, environmental), with particular reference to logistics, transport, and supply chain management activities.

What do best practice companies do differently in developing and implementing their own supplier sustainability compliance standards?

1. Best practice companies' auditing and control procedures are lean and primarily IT-based. Best practice companies do not delegate auditing, analysis or the development of solutions to consultants or other third parties. The people performing these functions, on-site, are often the firm's own employees and experts from relevant business units, who are well versed in the company's overall technology and operations. Hence these people are better able to identify and explore sustainability issues hand in hand with innovative solutions, instead of inevitably focusing just on compliance and "end-of-pipe" solutions.
2. They have a clear definition of the critical processes and critical suppliers based on thorough understanding of stakeholder requirements (see also sub-chapter four 'Step 3: Determine the Stakes – What is expected?').
3. They have a thorough understanding of existing and planned regulations and standards related to critical processes, and in consequence, they are able to define existing and future compliance gaps with precision.

4. They are able to translate compliance requirements into quantitative and qualitative indicators to assess critical processes and the related suppliers.
5. They have comprehensive decision criteria and action procedures for different audit results, and know clearly, for example, when to substitute a supplier and when not to.
6. They customize their standards with regard to country and partner-specific requirements, if absolutely necessary.
7. We have often seen that suppliers simply accept customer standards, as they usually have no other choice. But the feedback from practitioners shows that standards add real value to both sides, if the benefits are clearly communicated. Hence, best practice companies try to quantify and communicate the benefits of their standards, in order to motivate their suppliers to accept, apply, and invest (e.g. in training) in these standards (see also the Sharp case).
8. Many suppliers can tell of customer's audits and standards which simply ceased to be applied, without notification. These failures usually result from customers' unsustainable compliance projects which become irrelevant in the changing business environment. In contrast, best practice companies regularly modify and adapt their standards to changing conditions.

5.7 Conclusion

The following are some summary conclusions derived from the content of this chapter and from the bestLog research. They are mainly directed towards the strategic aspects of sustainable supply chain stakeholder management.

1. Managing the externalities of your organisation is an extremely complex process
2. The regulatory environment of supply chains has become highly complex and increasingly challenging for supply chain managers
3. Seven key action fields in secondary, especially regulatory supply chain stakeholder management were identified:
   (a) Think locally, act globally
   (b) Combine stakeholder communication with expertise and innovation
   (c) Communicate and involve
   (d) Collaborate and cooperate
   (e) Force a value-added, strategic and holistic approach
   (f) Eliminate Trade-offs
   (g) Track and monitor
4. Three major action fields in primary supply chain stakeholder management were identified:
   (a) Compliance and transparency in supply chain processes
   (b) Realising new efficiency potentials via process-optimisation and supply chain re-design

   (c) Product and service development and associated cost- and gain-sharing models
5. Four attributes to determine the importance of primary stakeholders were identified:
   (a) Their access to supply chain knowledge and information
   (b) Their substitutability
   (c) Their access to networks and infrastructure
   (d) Their value creation along the supply chain.
6. Two important criteria were chosen from the literature to classify supply chain stakeholders, in order to support a focused and differentiated action plan for sustainable stakeholder management:
   (a) A stakeholder's potential to threaten your organisation and your supply chain
   (b) A stakeholder's potential to cooperate with you
7. Generic stakeholder management strategies according to Savage introduced us to two important types of supply chain stakholder:
   (a) The "Mixed Blessing" stakeholder, ideally addressed with the "Collaborate" Strategy
   (b) The "Non-supportive" stakeholder, ideally addressed with the "Defend" Strategy

In the following references you will find again the most important recommendations based on best practice observations and analyses within the bestLog reasearch:

1. Best practice companies understand the standards imposed by their customers and comply with them. Furthermore they control their standards implementation effectively and report regularly to their customers as required.
2. Best practice companies "do good" and talk about it! They improve their reputation as reliable partners by demonstrating compliance with all relevant legal and industry standards in customer-facing processes.
3. Identify together with your customer and its other primary SC stakeholders further process fields in the end-to-end supply chain which may improve the overall efficiency of the supply chain. Think beyond your own and your customer's processes.
4. Looking beyond your own four walls for improvements requires that supply chain managers have adequate responsibilities and are offered effective incentives.
5. Identify the potential of state-of-the art ICT (Information and Communication Technology) to enable further efficiency improvements.
6. Rethink existing customer supply chain requirement and explore new win–win models together.
7. Check the price sensitivity of your customers and consider possible differentiation strategies for new or re-designed products, packages, and supply chain services, especially if you are a logistics service provider.
8. Emphasise collaborative research if customers are willing to explore potential new efficiencies, and if they require highly innovative products and services

9. Emphasise the early involvement of the customer's logistics and supply chain managers in collaborative product and packaging development.
10. Best practice companies maintain a global perspective and track developments, understand the differences between different regulatory requirements and standards, and gain knowledge of each country's and each sector's specific issues.
11. Best practice companies try to develop incentives and to motivate their suppliers to invest in complying with standards, especially their first-tier suppliers
12. Further, best practice companies motivate their first-tier suppliers to implement the same compliance culture and awareness, and to extend it to the second and subsequent tiers of the supply chain
13. Best practice companies verify the results. They are able to prove whether or not their suppliers are applying the given standards and laws in practice.
14. Best practice companies' auditing and control procedures are lean and primarily IT-based.
15. Best practice companies do not delegate auditing, analysis or the development of solutions to consultants or other third parties. The people performing these functions, on-site, are often the firm's own employees and experts from relevant the business units, who are well versed in the company's overall technology and operations.
16. They have a clear definition of the critical processes and critical suppliers based on thorough understanding of stakeholder requirements.
17. They have a thorough understanding of existing and planned regulations and standards related to critical processes, and in consequence, they are able to define existing and future compliance gaps with precision.
18. They are able to translate compliance requirements into quantitative and qualitative indicators to assess critical processes and the related suppliers.
19. They have comprehensive decision criteria and action procedures for different audit results, and know clearly, for example, when to substitute a supplier and when not to.
20. They customize their standards with regard to country and partner-specific requirements, if absolutely necessary.
21. Best practice companies try to quantify and communicate the benefits of their standards, in order to motivate their suppliers to accept, apply, and invest (e.g. in training) in these standards.
22. Best practice companies regularly modify and adapt their standards to changing conditions.

Tasks:

1) Base on Figures 5.2 and 5.4, list all stakeholders in your chosen supply chain and separate them into primary and secondary. Discuss their influence on strategy and supply chain strategy.
2) How you can classify your identified stakeholders (see Figure 5.7). What actions would you take to manage each group of stakeholders? Which strategy would you select and why?

3) Customers are one of your stakeholders; can you segment them into smaller groups according to their needs and influence? How do you communicate with each group? How can you establish long-term relationships?

Bibliography

Business and Poverty (2008): Innovative strategies for global CSR; how to develop global CSR strategies, manage risks and find new opportunities for your business in emerging and developing countries. Vienna: Global CSR Casebook.

Emmett, Stuart; Sood, Vivek (2010): Green supply chains. An action manifesto. Chichester: John Wiley.

Epstein, M.J. (2008): Making sustainability work. Best practices in managing and measuring corporate social, environmental and economic impacts. 1st ed. Sheffield: Greenleaf Publ.

Esty, D.C., Winston, A.S. (2009): Green to gold. How smart companies use environmental strategy to innovate, create value, and build competitive advantage. Hoboken, NJ: Wiley.

Freeman, R.E. (1994): Stakeholder theory of the modern corporation. In: Beauchamp, T.L., Bowie, N.E. (Ed.): Ethical theory and business. 4th ed. Englewood Cliffs, NJ: Prentice Hall, pp. 66–76.

Freeman, R.E., Harrison, J.S., Wicks, A.C. (2007): Managing for stakeholders. Survival, reputation, and success. New Haven: Yale Univ. Press.

Hall, J. (2006): Environmental supply chain innovation. In: Sarkis, J. (Ed.): Greening the supply chain. Berlin: Springer, pp. 233–249.

McKinnon, Alan C (ed) (2010): Green logistics. Improving the environmental sustainability of logistics. London: Kogan Page.

Mitchell, R.K. (1997): Towards a theory of Stakeholder identification and salience: defining the principle of who and what really counts. In: Academy of Management Review, Vol. 22, No. 4, p. 869.

Piotrowicz, W. (2008): Designing Closed-loop Supply Chain: Implementation of the WEEE directive in the UK, The European Retail Digest 56:12–15

Polonsky, M.J. (1995): A stakeholder theory approach to designing environmental marketing strategy. In: Journal of Business & Industrial Marketing, Vol. 10, No. 3, pp. 29–46

Porter, M.E., van der Linde, C. (2008): Green and competitive. Ending the stalemate. In: Harvard business review on profiting from green business, pp. 121–134.

Sarkis, Joseph (ed) (2006): Greening the Supply Chain. London: Springer.

Savage, G.T., Nix, T.W., Whitehead, C.J., Blair, J.D. (1991): Strategies for assessing and managing organizational stakeholders. In: Academy of Management Executives, Vol. 5, No. 2, pp. 61–75

Suchman, M.C. (1995): Managing Legitimacy: Strategic and Institutional Approaches. In: Academy of Management Review, Vol. 20, pp. 571–610.

Thomlison, B.M. (1992): Environmental consumerism must meet needs of corporate stakeholders. In: Marketing News, Vol. 26, No. 9, p. 12

Waters, D. (2007): Global logistics. New directions in supply chain management. 5th ed. London: Kogan Page.

Chapter 6
Outside Your Control and Influence: Managing the Unexpected

Graham A. Ewer

Learning Goals.
By reading this chapter, you will:

- Learn about the nature of risk in supply chains and where it fits in when conducting sustainable business
- Learn how to manage risk inside your business and your immediate supply chain
- Learn how to manage risk from outside your business, including how to cooperate with others to reduce supply chain risks
- Learn how to reduce the risk to your supply chain of the unexpected beyond your control
- Learn about the important contribution of best practice to reducing supply chain risk

6.1 What's This All About?

In the earlier chapters, this book has dealt with supply chain operations and performance: this chapter deals with the threats to successful supply chain operations. Few successful enterprises are without uncertainty and the unexpected – what we commonly call *Risk*. Indeed, effective risk taking is at the very heart of successful business: few over cautious enterprises are agile enough to survive in today's marketplace. But, much risk is unforeseen and prone to present unusual challenges that are difficult to deal with, and seldom timely (Fig. 6.3).

Managing risk in *end to end supply chains* is especially vital: all of them involve a collective interdependency in some form or another. It is in the nature of things that the extent of this interdependency is not always apparent – many people can only see, and influence, activities that are their responsibility, without appreciating the context of what they are doing in the larger end to end supply chain in which they play a part. This increases the possibility of the unforeseen and the unexpected: it needs the special attention that we consider later in this chapter.

C. Tyssen et al., *Sustainable Supply Chain Management*,
DOI 10.1007/978-3-642-12023-7_6, © Springer-Verlag Berlin Heidelberg 2011

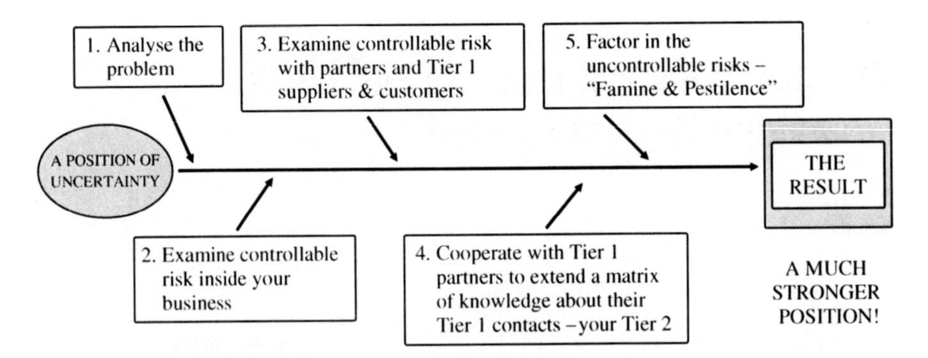

Fig. 6.1 A roadmap to reducing risk

What are we trying to do? Let us start by saying what we are not trying to do! First and foremost, there is no template or formulaic solution to all of this and this chapter does not set out to provide one. Second, this chapter is aimed primarily at those whose task is to pilot and operate a logistics business. Its principles will be as useful a guide to practitioners in large multinational companies, all of whom are making a substantial daily investment in risk analysis, as it will be to practitioners in Small and Medium Sized Enterprises (SMEs) who are usually short of resources and proportionately at higher risk as a result. This is not to say that it is intended for practitioners alone: it will be useful too as a guide for all those engaged in teaching good practice in logistics. In an effort to bridge this spread of interests, this chapter examines the nature of risk from the perspective of a medium size business through the eyes of its owners and managers. It sets out to address a number of principles, processes, and tools which can be applied generally so that all businesses are better equipped with the resilience to prosper in an uncertain and challenging market place.

Why? Although the reasons for reducing risk in our businesses may seem obvious, it is still worth asking why we should spend time on risk. No reliable statistics seem to exist as to how many businesses fail as a result of the impact of risk on their supply chains although the evidence of failure is only too common. On the grounds of statistical distribution alone, a substantial proportion of those who fail will do so because of supply chain failure of some kind. The lack of statistics is hardly surprising: the field is broad and many enterprises fail because of invisible failure in other peoples supply chains – SMEs are particularly vulnerable to losing business and failing because of mistakes in other peoples' supply chain practice beyond their control, but of which they are a part. All this provides convincing evidence as to why we should try to reduce risk. Identifying and mitigating risks reduces the chance of this happening to your business. The process will improve your business also, and give you another mark of quality to hold up when negotiating with clients and suppliers alike. It brings with it substantial indirect advantages in that by doing business better the sustainability of your supply chain can improve. It is part of knowing and demonstrating your capabilities, and it is core to effective business. Even a cursory glance at the cases of best practice investigated by *bestLog* indicates the importance of doing things well as a means of minimising risk. A strategy for risk management is an essential part of your overall business strategy and it is a necessity – in both good times and bad.

Who should read this? This chapter is relevant to all those involved in managing supply chains, and to those who teach good practice in this field. It is especially relevant to SMEs, and not only to those businesses engaged directly in the logistics industry and supply chain management either. The principles and practices covered here can be applied to more general business with considerable benefit.

How is this Chapter Organised? This diagram illustrates how this Chapter is organised (Fig. 6.2):

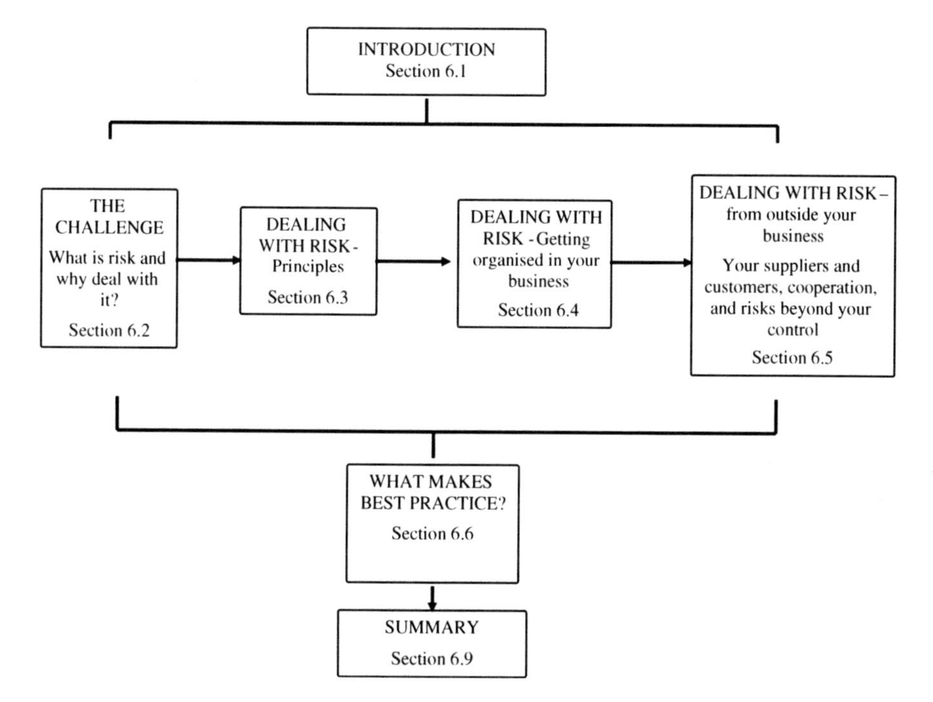

Fig. 6.2 How the chapter is organised

6.2 The Challenge: What Is Risk and Why Deal with It?

> Risk – "a situation involving exposure to danger > the possibility that something unpleasant will happen" – Oxford English Dictionary

Putting risk in perspective. As a starting point, it is important to put risk firmly into perspective before attempting to address how best practice, or good practice, can help to reduce it.

- It is commonsense that nothing is without risk, and that nothing is gained without taking some risk. A risk free solution is an illusion and every attempt to minimise risk has a price which has to be assessed and paid.
- The penalties of failing to take measures to manage risk are self evident – they hasten business failure.
- The management of risk is central to overall business strategy and must be relevant to it: some businesses are more risk averse than others.
- The management of risk in supply chains should be seen as a subset of overall risk management strategy in a business: it must be directly relevant to the latter, but may differ in emphasis – for example, a business that is not averse to risk overall might seek to minimise risk in its supply chain to provide strength to take risk elsewhere.
- Like it or not, the management of risk is a constant in both good times and bad. It is in the nature of things that to be sucessful, the conduct of risk management must stay in step with the aims of the business and the risks to that business at any particular time.
- In the final analysis, risk is always present and whilst the approach to managing it is a constant, the judgements taken as a result of this must always be relevant to the business and the environment it operates in.

This is the essential backdrop against which we can consider how to handle risk in supply chains (Fig. 6.3).

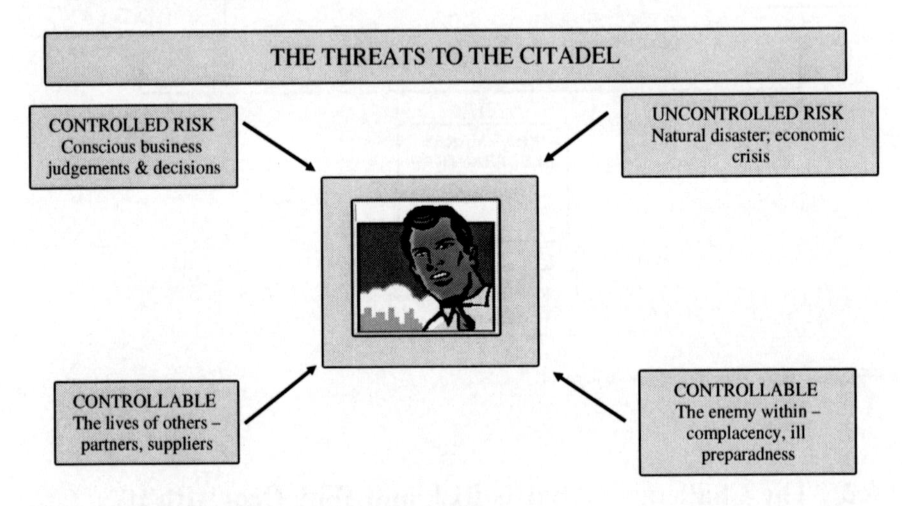

Fig. 6.3 Threats to the Citadel – the nature of risk

Risks come in all shapes and sizes, but for easy reference can be categorized broadly as Controlled, Uncontrolled and Controllable:

Controlled risk: Risk taking is an essential part of business management – even the smallest enterprise develops a business plan of some kind that weighs up strategies and practices matching resources to tasks with a degree of calculated risks that underpin profitability. For the most part, these risks are controlled in the sense that they have been evaluated and built into planning and activity. For the purposes of this chapter, we will do no more than acknowledge this process as part of normal business, referring to it only as necessary. For our purposes, it is the uncontrolled risks, and the controllable risks that are of the greatest interest: the former because of their unpredictable and potentially damaging nature; and the latter because of their potential, if addressed, to protect against the unexpected, and to reinforce the effectiveness of normal business.

Uncontrolled risk: If asked, most people will identify risk as being synonymous with uncontrolled risk in the shape of natural disasters, emergencies or economic crises on a national or global scale that are beyond their ability to control. They take many forms and supply chains are particularly vulnerable to their consequences because supply chain activity relies heavily upon satisfactory infrastructure and pre-conditions, such as secure global trading. No enterprise can afford to become fully equipped to deal with every eventuality and the means of dealing with them often lie in the hands of others, (and some might say, Providence). But, uncontrolled risks can be mitigated at least to some degree: we deal with this later.

Controllable risks: Internal and External. Yet many risks in the supply chains can be reduced, and even converted into controlled risks because they can be recognised and addressed. In principle, controllable risks divide into those arising within the business – internal risk, for example, bad internal supply chain practice – and those arising from the conduct of others working with you – external risk, for example, the conduct of your suppliers and customers. Both varieties are usually within reach and can be tackled at reasonable cost by the application of good practice. It is here that the bestLog project seeks to provide material which can be applied to improve supply chain performance and resilience at every level. But, before we get involved in the detail, having understood the widespread nature of risk, we ought to have some idea of its extent and significance for us right now.

How real is risk today? Having discussed risks in the abstract, it is important to focus upon the practical. We must consider whether supply chains are more vulnerable today: is risk increasing, and are risks of various kinds becoming more widespread? Certainly, there are some causes for concern:

- *Leaner supply chains*: In the last 20 years there has been pressure to reduce cost by reducing the level of inventory in supply chains as a whole. These leaner supply chains are more vulnerable to unexpected disruption.
- *Reliance on transport*: This process of reducing inventory in leaner supply chains has resulted in much greater reliance upon transportation to provide responsive supply chains.
- *Global outsourcing*: Further more, globally, the practice of outsourcing manufacture, has greatly increased reliance on transportation, compounding the pressure to substitute transport for inventory in leaner supply chains.

- *Infrastructure*: Transportation depends on a steady infrastructure in the widest sense. It relies upon reasonable fuel price, secure transit conditions without interference from, for example, terrorism, and needs space to operate effeciently, for example, an uncongested road or rail network. Overall, transportation is very vulnerable to disruption in a variety of forms, affecting supply chain performance accordingly.
- *Sustainability and environmental issues*: The impact of environmental issues, be that physical in the shape of climate unpredictability or additional cost, is increasing. Increasingly, customers include judgments on environmental performance – how you tackle these issues – alongside business performance.
- *Social issues*: As with environmental matters, so social issues – safe practice, consideration for employees – have become a more closely regarded element of overall performance which, if not taken into account, become a source of risk.
- *Technology*: Technology has become core to the operation of supply chains and, as reliance on it increases, with this grows vulnerability due to, for example, over pressure on the World Wide Web/Internet, the challenges of information security, and the lack of human skills to handle these challenges.
- *Economic stabilty*. Economic stabilty has a key influence on supply chain risk. At the strategic level, the practice in economic good times of relying upon a lean inventory, or moving production to economies where production is cheaper, become questionable when economic conditions are less favourable. Economic instability brings its own brand of business risks, especially in terms of confidence and trust in cooperation and funding for investment in improvement. Moreover, such has been its effect upon the economics of business that previous assumptions, such as, and the assumed benefits of outsourcing production globally, are called into question.
- *Europe and Asia*. From a European perspective, the emerging Asian market and strength of Asian manufacturing produces a unique strategic risk in this context. Not only is production a long way away, but costs are variable with a limited range of alternatives resulting in complicated supply chains and routes to market.

Of course, for each of these, there are offsets – for example, technology has been the key to greater efficiency, in part perhaps outweighing its risks. In reality the threat facing us is more in the nature of a matrix where some issues stand out for a time before giving way to others. Nevertheless, our empirical bestLog experience, and a number of authoritative business confidence surveys,[1] all indicate that strategic risks, many of an uncontrolled nature, are increasing. This gives us all the more reason to look very hard at how we can reduce the risks – to do so, let's look first at the principles involved.

---

[1]For example, Research Series, published annually 1988–2008, "Logistic Trends and Strategies" Bundesvereinigung Logistik (BVL) and Berlin Institute of Technology Logistic Department (TU Berlin).

Case:
Famosa – Full Speed Supply Chain
Using IT and scheduling to reduce risk in a complex supply chain
(See Case Collection in Part IV of this book)

6.3 Dealing with Risk: Principles

This section covers the underlying issues that are the foundation of the way in which we deal with risk. These vary from our basic attitude to taking risks – for example, an experienced mountaineer's attitude to risk when climbing is quite different to that of an untrained novice – to the important matters of leadership. These are the constants which together form the common denominators that go with any solution. When matched with appropriate technical solutions, they provide the means of mitigating or relieving risk.

The Psychology of Risk. There have been a number of important studies into supply chain vulnerability and the management of risk – some are listed in the supporting reading list. One of these makes two important points[2] that still hold true today:

> Managers with supply chain responsibilities focus, in general on internal operational risks. They are not explicitly required to address supply chain vulnerability or resilience. Business continuity planning tends to focus on the internal network yet the message that needs to be understood and acted upon is that the biggest risk to business continuity may well come from the wider supply chain rather than from within the business.

Put another way, most of us are good at addressing the obvious risks close to home, but pay far too little attention to the wider supply chain of which we form a part. BestLog's work suggests that processes to address risk internally are available, robust and applied.[3] But, too few managers look beyond their immediate suppliers and customers, and are unaware of trends and weaknesses elsewhere as a result. Smaller and medium size businesses (SMEs) will often say that they have a genuine difficulty in finding the resources to do so. Yet, in so many ways this is a cultural and training issue rather than one of resources. Our evidence is clear that best practice involves:

- Understanding where you stand – a clear understanding of one's place in the overall supply chain.
- Knowing your customers and suppliers – a clear knowledge and understanding of not only immediate suppliers and customers, but of those with whom in their turn they work also.

[2]Creating Resilient Supply Chains UK Department of Transport and Cranfield University School of Management 2003.

[3]For example, the SCOR Model and business standards such as ISO 9000.

- Knowing the trends – a lively interest in global trends and
- Facing facts – a firm determination to avoid the common human pitfall of trying to hope that potential problems will go away!

Supply chain managers who can match these criteria can make a better plan to overcome risk. It is as much an attitude of mind as it is one of education – there is no room for complacency or inattention. Rather, it is the business of developing a risk culture that is neither *risk averse* nor risk pre-occupied, but rather is *risk alert*. But, to be risk alert, you must know what you are looking for.

Recognising risk when you see it. As an extension of this, bestLog's research, which examined the issue of best practice from an economic, environmental and social perspective – see Part I – has demonstrated very clearly the importance of recognising the constantly changing nature of risk. It is in the very nature of uncertainty that the weighting to a particular risk will come and go and former strengths may transform into weaknesses with the passage of time. Equally some risks are less obvious than others – for example, new social legislation impacts upon practice and, if it is overlooked, this becomes a source of risk to your business. The recognition of risk needs knowledge and the commitment of leadership.

Leadership. A proactive risk alert culture will not grow by itself. It demands leadership from the top. Top management needs to understand the nature of risk and where it stands in relation to their business. Above all they must be committed to a risk culture and spread its key tenets through their organisation so it can benefit from the initiative that grows from understanding. It also means listening to your people and benefitting from their full engagement. Some enterprises we have seen have developed a dedicated multi-discipline risk management team within their organisation charged with a regularly updated risk register considered regularly at board level. Whilst perhaps this need not be so formulated in an SME, the principle of this approach can still be employed to very good effect.

Visibility. But, even given such leadership, such a team will only be as good as the quality of the information it can obtain as a basis for its deductions – in other words its visibility of the supply chain. Visibility in this context goes beyond understanding performance internally, or even with one's immediate suppliers and customers. It extends to information up and down the end to end supply chain in the necessary detail to identify likely risks that might arise. Often this is strictly limited, sometimes through lack of effort or a lack of understanding of the value of such information, but more usually because such information is hard to obtain in a commercially competitive environment.

Collaboration. In the end, this can only be solved by collaboration of some kind and there is growing evidence that collaboration is increasing, and so is a better understanding of its value. However, whilst few would argue against the necessity and value of due diligence during contract brokerage and acquisitions, voluntary collaboration to avoid risk is viewed with more suspicion. The supply chain industry has a history of adversarial relationships and some courage is needed to alter this culture to one of more cooperation. But, all the evidence suggests that best practice to manage risk lies in greater collaboration between participants. This takes time to establish, for there no one size fits all solution, save that simplicity has great merit. Usually, the most effective collaborations have been between those enterprises which have cooperated in an evolutionary way – making a few simple things work first to build confidence and value based on experience.

Information, knowledge and measurement. As with many things, these matters of principle are vulnerable to the indiscipline of detail. Their success as a means of reducing risks depends upon a common understanding of the information upon which to base implementation. This common language extends to technical compatibility, common measurement, and a common set of tools with which to work. Moreover, different types of information are needed at different levels: at the strategic level, informed qualative assessment of trends has its place; at the operational level, much tauter quantitive measurement is needed.

Know your business. All this pre-supposes a foundation of knowledge about how supply chains work, and how one's particular business works. This fundamental understanding is a critical foundation for effective risk management, both as a basis for judgment and as the source of regulation for performance and change.

A clear understanding of these principles is the essential basis for turning to the practical challenge – "how to reduce risk in my circumstances?" Bear them in the back of your mind as we turn to just that question.

6.4 Dealing with Risk: Getting Organised In Your Business

This section deals with the assessment and management of internal controllable risk – how to assess and manage risk in your business on the foundation of the principles considered above. The diagram below holds the key to success – a continual process of improvement (Fig. 6.4).

In practice, it is hard to identify any enterprises that are unaware of risk in some way or another. But, we all know that some handle risk better than others, and are more effective and confident in the marketplace as a result. When you examine these successful players, a common factor is evident: in each case they approach the issue in an organised way, analysing their situation, assessing what is needed, and putting in place a plan for action. From the outset, it must be realised that the management of risk must be approached in a systematic and holistic way. A step by step approach seems to pay off for many who embark upon this in earnest;[4] but in any event a robust and easily understood system is essential, defining responsibilities, tasks, and implementation.

Getting organised to deal with risk is fundamental. There is no "one size fits all" template, but there are some points that cannot be ignored:

- *This is about your business*: Whilst the examples set by others can, and will, be helpful, that is only insofar as you can apply them. What works for you must take

[4]There are a number of available bestLog case examples.

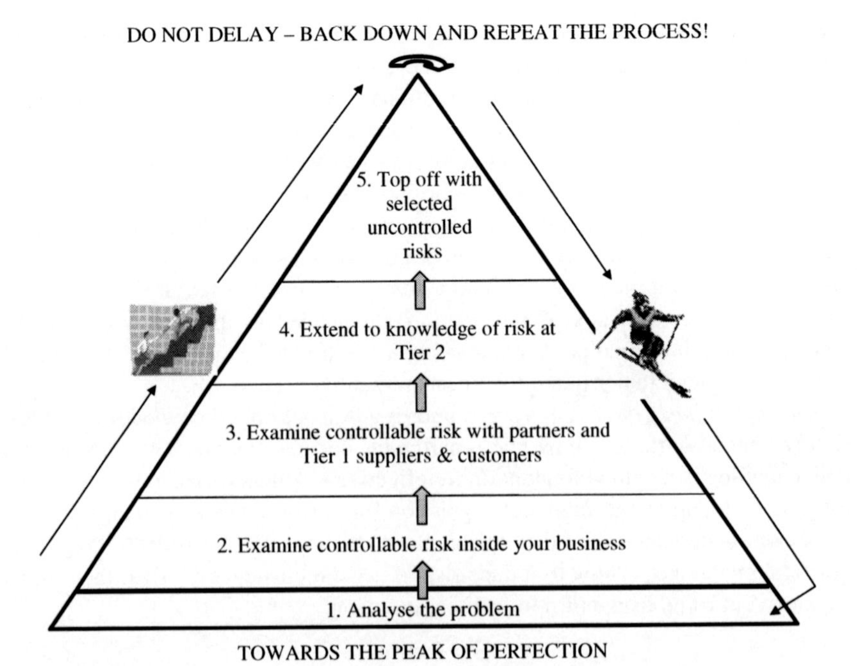

Fig. 6.4 Towards the peak of perfection – a process of continuous risk assessment

into account your resources, your needs and your position in business. Put simply, as a first priority you need to understand where your business is positioned, and what is likely to be proportionate in your response to risk. For example, in a small business, it might be most appropriate for the owner to simply wrap up supply chain risk with his other responsibilities for his business. In a slightly larger one, a single director might take responsibility, and so on. In some senses, the most important thing is that a definite decision has been made to address supply chain risk and how in general to get organised.

- *The foundation role of supply chain knowledge*: You cannot do this without the foundation of a good understanding of how supply chains work as the basis for judgment and decisions.
- *What is missing?*: If such a decision has not been made before, it will usually be the case that something seems to be missing – information or the right skills in the business, for example. Above all, circumstances can look chaotic: some means of dividing the problem into manageable portions is essential.
- *Business Analysis*: The business world is awash with methods of business analysis, all of which have their merits. There is a wide choice of consultants offering solutions. But, for most SMEs this issue is one of resource and we suggest that initially much can be achieved by a rather more basic approach to supply chain risk analysis by focusing your assessment into the three broad

areas: supply chain risks inside the business; supply chain risk in your immediate circle of Tier 1 customers and suppliers; and, the risks from the world at large.

- *Start at home*: The first place to start is to look within your own business. Not only is this the easiest step to make, but it is, or should be, what you know most about, and it is the area you can most easily influence things directly. Moreover, in one set of actions you can both organise for risk and begin to identify risk.
- *Tier 1*: Dealing with your neighbours, your immediate customers and suppliers, is more difficult, but will follow naturally from any internal risk analysis.
- *Gaining momentum*: The process gains a momentum of its own once begun and begins to throw light not only on your supply chain risk, but also on how you need to organise your approach to your supply chain risk in particular and
- *Recording risk – the risk register*: Central to all this lies the discipline of registering risk and a risk register. Too many enterprises have plenty of risk information that is neither accessible nor managed. The risk register is the missing link in these circumstances which must not be a sterile template, remote from the needs of the business. On the contrary, an effective risk register is the focus for the collection of risk information that is core to the business – a live document to which all refer when making decisions that are central to the business.

Making the best use of knowledge and education in your business. A good understanding and knowledge of supply chains is fundamental, yet many underestimate its importance. It is seldom very long before the process of getting organised highlights the importance of knowing your business, and the breadth of professional knowledge that is necessary for managers to make informed judgements and decisions. Put simply, if you do not have a good knowledge of the practice of supply chains, and of such matters as how to measure performance, how do you know the extent of the risk that you carry? It can be depressing to look at something that you already thought was good from a fresh perspective, only to find that some of the remedies are going to take longer than you thought. But, the issues have to be faced: there are solutions aplenty and time can be found.

Knowledge is the key and it has enormous collateral benefits:

- *Educated professionals*: Supply chain education is fundamental. Evidence suggests that whilst standards of specialised knowledge in the logistics industry generally are quite good, broader knowledge of supply chain management is not good. For example, transport operators are operating successfully within the parameters of their specific function and running safe fleets efficiently. But, too often these same managers have less knowledge of supply chain management and less understanding than they should have of their position in the supply chain and their vulnerability as a result. This is a source of risk in itself, whilst the managers themselves are not well placed to analyse the risks in their enterprise either;
- *Barriers*: Education is, of course, the answer, but there are practical barriers in terms of time – people have to find time to learn – and in terms of motivation – some of us do not like to admit that we have got to "go back to school"; and,

- *Broader benefits*: The collateral benefits of better educated managers should not be underestimated for they extend well beyond risk management. They make an important contribution to commercial agility and profitability as well as providing a vital foundation for the means of dealing with the unexpected.

Despite the supposed difficulties, the arguments for improving the knowledge of supply chain management in your company are compelling and there are some things that can help:

- *The unique needs of your business*: Keep firmly in mind the real needs of your business. For example, in many SMEs, the need is for a greater breadth of knowledge rather than depth. Breadth provides better understanding and the knowledge of where to look when greater depth might be needed. Courses are available throughout Europe designed to provide this: bestLog maintains a European database of such information – http://www.bestlog.org;
- *Recruiting the right talent*: Examine your recruitment policy carefully and make sure that you look for breadth of knowledge in the managers you recruit in future.
- *Education programmes*: Consider closely whether some in house education programmeme might have a place and do not be conservative when it comes to how it is delivered – the educators might come to you, education might be part time, by correspondence course or e-learning.
- *Finding the time*: Somehow, find the resources to educate key players in your company, and maintain the continuity of such investment. Margins are always tight and the return on investment of an intangible such as education hard to measure. But, although the benefits of better knowledge take time to work through, work through they will. Always track improvement and ask the question "Would we have done this if we did not know how to, and how much benefit have we gained as a result?" It is surprising then how you can see what education may have realised for you.
- *Human nature*: In a busy company, finding the time and the resources is only half the problem. Human nature is the other half: your team has to be persuaded of the value of education. Above all, this is a matter of leadership to create this understanding and build on the benefits for all that is inherent and
- *"Training is a waste of time and money"*: Finally, there is still an unspoken perception abroad that educating and training people is a waste of time because once trained, your people will leave, or be headhunted for better jobs, thus adding to risks just when you were trying to do something about it. There is more than a grain of truth in this perception. But the converse – becoming a progressive employer who trains his people, profits accordingly, and attracts good people accordingly – usually does not get a mention! Investment in people breeds loyalty.

All this boils down to commonsense investment in education to ensure that your enterprise can implement the best possible judgements and decisions to address internal risk.

The key role of efficient internal business. The odds are that if we strive for doing the most efficient business, we also contribute to reducing the risks both to our business and in our supply chain. By its nature, an efficient supply chain using best practice in the way best suited to the business will be more robust and sustainable, as well as having a commercial edge. But how does an SME measure and test its supply chain and ensure it is fit for purpose?

There are certainly plenty of toolkits around to do so. The bestLog project has found a variety of toolkits which can be used and there are a number of accessible pieces of research designed expressly to assist SMEs.[5] For example, work sponsored by the UK Department of Transport with Cranfield University and LCP Consulting in 2003 provides both a practical guide and a self assessment work book for companies. Although the choice is wide, it is also potentially confusing. Some simple points are worth bearing in mind:

- *Cost vs. benefit analysis*: As a first essential, always look closely at the balance between the benefits to be gained and the cost of achieving them in the context of your business.
- *Be open-minded as you do so*: We have all met the enthusiast who advocates his or her system in its entirety with near religious fervour to the exclusion of all others. But for the smaller business the answer may be parts of one or more tool kits targeted with care with little cost or complexity using your current in house resources.
- *Outside help*: Some external advice may be helpful too provided that you are absolutely clear on your precise objectives and the cost versus benefit equation involved. Consider too whether a customer or supplier might be a source of advice. After all, they stand to benefit directly from your supply chain improvement, and might themselves benefit from reducing risk in their own supply chain elsewhere.
- *Keep it simple and proportionate*: For most smaller enterprises, the ubiquitous spreadsheet and commonly available software is more than sufficient to carry out what is needed. It should be your information that is priceless, not your system!
- *Future compatibility*: But compatibility is an issue, and with an eye to the future, look to what systems are in use by your customers and suppliers. As things develop, you will want to share information.

In sum, there are plenty of tools around to measure and improve your supply chain practice: the key is selecting the right toolkit for your business in particular, and resisting the siren voices of complexity!

Identifying where your internal business practice is most vulnerable: internal business risk hotspots. Once you are organised for risk, you will be turning swiftly to analysing risk. There is a temptation to either see so many risks that it is hard to know where to act first, or to be mesmerised by a single risk to your supply chain to the exclusion of all else. But, in practice, there are some potential hotspots that are

[5] A selection is shown in the "Further Reading" for this chapter.

common to most businesses and it is useful to have in mind some of these as a framework for risk analysis. In each case, it is best to start with the straightforward before getting drawn into the more complex. On each occasion it is a matter of: what is necessary; what may be missing; if so, what might be the effect; and, the cost of fixing it. Common areas for consideration include:

- *People*: Have I got enough of the right people? Start with the simple – the qualified people you must have, for example, a fire safety trained employee required by law, before working up to more sophisticated issues related to the quality and knowledge of your people in the context of reducing supply chain risk by improving performance.
- *Buildings*: Across the piece from basic matters like insurance to alternative accommodation if there is, say, a disastrous fire.
- *Operational equipment*: From the simple questions – "Is it insured?" – to the more complex questions – "Is it the right equipment?"
- *Information Technology and data security*: Examine both process and physical equipment, from simple things like back up rules and the arrangements for IT support, to more difficult issues such as whether your IT matches the most effective supply chain processes you want and
- *Systems*: Are they well understood? Is there a risk conscious culture?
- *Legislation*: Is there a system in place to track legislation and to prevent oversight leading to risk?

Risk Registers. All these considerations involve quite a lot of detail, which can become unmanageable and even conceal important risks. This detail needs bringing together in a manageable way in the form of a risk register. Unlike much of the practice of risk management, the use of a risk register as a form of template or check list can be helpful, particularly to prevent oversight of perennial risks which are low level, but potentially very damaging if they are missed – for example, renewing one's fire insurance. But, they can do much more than this if well conceived, acting as a record to which all can refer. The best risk registers have 2 distinguishing features – they have been drawn up top down by the business so they reflect the strategic priorities of the business, and they are owned in the fullest sense by an appropriate senior manager responsible for the efficiency of that business. But, even when this is the case, beware of being lulled into a sense of false security – your risk register is only as good as the strength of purpose to keep it up to date! An example of the framework of a risk register for an SME is attached at the end of this Chapter.

Case:
BT – Improving Efficiencies Within a Scheduled Trunking Service
Improving usage within a scheduled trunking service to reduce carbon footprint and improve efficiency.

(See Case Collection in Part IV of this book)

It will not be long before work on internal risk draws you inexorably to the risks from without – initially with immediate suppliers and customers, and subsequently further afield.

6.5 Dealing with Risk from Outside Your Business

The principles and practices we have discussed when dealing with our internal business are equally applicable to dealing with risks from outside the business, but they have to be tackled on a wider and more complicated playing field. By definition, one has less direct control and much greater reliance upon persuasion and complex relationships in a competitive setting. This section deals with these aspects that particularly characterize this situation – those with whom you must deal, how to do so, and the vital role of cooperation.

Your immediate customers and suppliers. The supply chain industry as a whole, and many other parts of business too for that matter, have something of a tradition of adversarial relationships, clouded by a grudging recognition of mutual dependency within a more distant end to end supply chain. But all of us acknowledge, perhaps not too vocally sometimes, our dependence upon our immediate first tier customers and suppliers. Usually, we do not enter business with them without simple checks – classically the bank reference and informal research into reputation. They are the obvious place to start when looking beyond internal risks. But, even amongst the closest partners, this is not an easy issue:

- *Commercially sensitive information*: Sharing a desire to reduce risk is one thing, but this always involves information, and sharing information which is commercially sensitive is another matter all together.
- *Realism*: All parties involved need a strong sense of proportion and clarity about what is to be achieved, why and how.
- *Trust*: All this demands trust which needs nurturing and time to come to fruition and
- *Data exchange*: The difficulties of achieving a common understanding of data and a common currency for the exchange of data should not be underestimated. Often small company systems are incompatible without some investment and locally developed definitions of process can hinder cooperation.

But cooperation is not rocket science either. The difficult part is often starting it rather than doing it once you are under way. Overtures have to be carefully prepared: all experience shows that for SMEs evolution is the method most likely to succeed. There are few overarching solutions in any case, and building progressively by cooperating on small things that can produce results builds the confidence and trust that is needed. Good cooperation is a fundamental.

Case:
Wincanton – Return Logistics, Implementation of WEEE Directive
Supply chain improvement to environmental advantage, majoring strongly on cooperation
(Piotrowicz (2008) Collection in Part IV of this book)

Cooperation as a way of reducing risk. Given the importance of cooperation, then it necessary to look at how to proceed with cooperation. The key to successful cooperation lies in a robust and repeated process, shown in this diagram (Fig. 6.5):

- *The case to cooperate*. First, examine the case for cooperation – the benefits and the disadvantages of doing so;
- *Potential partners*: Next, assess your potential partners against a matrix of their value to you in reducing risk set against the likelihood that you can achieve effective cooperation. In some cases, an easier option may be to partner with a lesser supplier or customer to achieve results which may be more useful in persuading other more valuable candidates into the fold. In other cases, it may be more sensible to aim higher and get an influential potential partner involved at an early stage.
- *Make a plan*: In any event, before making a move, work up a plan and a case, viewed as far as possible from the potential partner's point of view. The case must include the ground rules of a code of conduct and is designed to be the starting framework upon which to build an effective relationship.

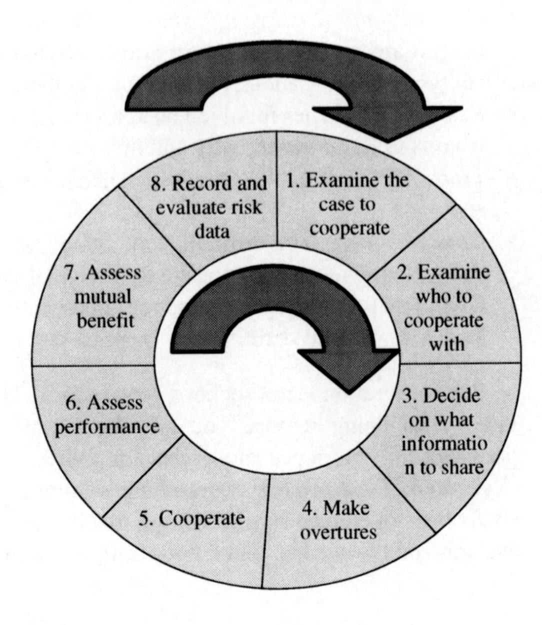

Fig. 6.5 A virtuous circle: Assess > Act > Measure > Record, then back round the process

- *Disclosure*: Be frank, and recognise that you must be prepared to disclose information if things are going to work. Decide how much before you start, not when you are under way!
- *The approach*: Tailor your approach with care using all you know about your customer or supplier. This not only ensures that you choose the right partner, but also impresses your would be partner of the seriousness of your intent.
- *Continuity*: Be clear about the cost of continuity – this will take work to get off the ground, and to keep it aloft, and it will need participants that are confident in their authority to adapt to circumstances as cooperation develops.
- *Be open*: As soon as it is practical, be open about the work in hand to prevent misunderstanding, rumour, and opposition through ignorance and
- *Protocols*: Once the ice is broken, early decisions are needed on the protocols of cooperation.

The ultimate aim must be to get risk reduction cooperation into the normal process of business between the partners with demonstrable benefit to those concerned that others can recognise with a view to participating as well (Fig. 6.6).

Getting a view of the end to end supply chain. It will not be long before the cooperation between first tier customers and suppliers opens up the prospect of a better understanding of the risk inherent through their respective second tier customers with some information emerging quite early in the initial phase. The benefits of reaching out to the second tier and further are likely to be significant on two counts:

- Less risk and better business through supply chain improvement and
- Better risk control because greater knowledge decreases the risk of the hitherto invisible unforeseen occurrence.

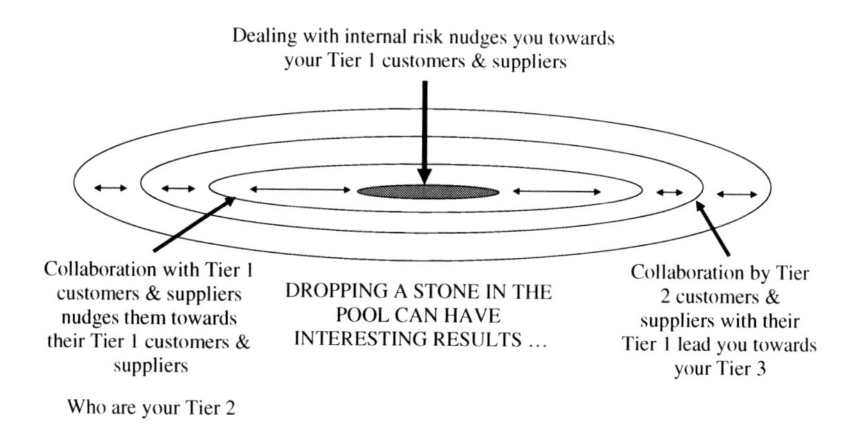

Fig. 6.6 Dropping a stone in the pool can have interesting results

However, there are penalties as well:

- Spreading cooperation takes time as first tier partners spread their efforts to others and a network is constructed. For medium size companies such a network can become quite complicated and
- It can also be quite costly, and it is vulnerable to growing a self defeating bureaucracy unless there is ruthless attention to simplicity and identified results.

The fruits of cooperation are not the only way to get an overview of the end to end supply chain and your place in it. There are analogies to be drawn with military intelligence which uses a number of sources to draw up a picture of the battlefield. In the same way, and without too much investment, commerce can organise itself to draw up a picture of the risks which may lie out there in the broader supply chain. The sources are generally open, and may be of variable quality, but they are open to sensible interpretation provided you are organised to do so. This process does not involve a whole department to do it – for example, in a small company this may be a matter of knowing something about the other first tier relationships your immediate customers and suppliers have and monitoring the publicly available financial performance of those involved. You may do this anyway. The essence of its value in this instance lies in the discipline of including it in the risk assessment and risk register process so that it is available to consider regularly.

This is very important, for the secret of risk mitigation in a wider context, lies in marshalling information with a solid regime of assessment. Through a risk register, you can combine what you know from cooperation with what you can find out organised in such a way that judgements can be made. Such a way of ordering your judgements in a common framework is especially important when you come to the more nebulous uncontrolled risks that surround us.

6.6 Uncontrollable Risks and How to Handle Them

Like it or not, some risks are going to be beyond our control. Most often, the best we can reasonably expect to achieve to mitigate these risks is have some idea of how likely an event might be, and what sort of general plan we might have to deal with it. This is an area fraught with uncertainty and deeply prone both to misunderstanding and the perversity of circumstances – "usually the least expected turns up". Nowhere is a sense of proportion more necessary to capture a realistic range of risks that are truly relevant to the business. In many cases, at the catastrophic end of the scale everybody will be in the same position and living on their wits. The most sensible approach here is probably to do no more than to decide and record that the risk mitigation plan in these circumstances is to deal with whatever turns up at the time!

Most of the risks are familiar and exposed through the media. They include:

- *Natural disaster* at a large scale which disrupts normal life
- *Economic difficulties* at a large scale which slow commerce
- *Security difficulties* which disrupt both daily life and trade

Only the broadest contingency plans can do much to help us with these. They may amount to no more than noting potential collaborators should the unexpected occur.

But, rather more important and relevant is the issue of trend spotting and tracking – identifying those developments that are likely to become a threat to business. The irony is that most of us do this anyway to some degree, but may not bring the results to the table as identified risk. That is easy enough to do by means of the risk register if a few key fields are monitored:

- Supply chain practice – developments in supply chain practice which might overtake us if we fail to identify them.
- Data handling – similar developments in data handling, processing and transfer.
- Global security – developments which may impose on our business – new security procedures, additional cost.
- Climate change – pointers to the future.
- Environmental and sustainability developments: fresh initiatives to reflect environmental concerns in supply chain practice and
- Social developments – new emphases which might affect us.

This is not an exhaustive list – rather, the purpose is to develop a monitoring net which, by identifying a limited number of key areas for over-watch, has a reasonable chance of catching not only the trends that are listed, but also others of a like kind which might arise at the same time. For example, monitoring the data handling field has a reasonable chance of alerting us to fuel related risk because data handling needs power and power generation needs fuel. Rough and ready this approach may be, but it is about the best that can be done at SME level without disproportionate cost. Precisely how to monitor these risks depends very much on the enterprise in question. In a very small business this might be undertaken informally by the owner or managing director, who might also own the risk register. In a medium sized enterprise, it seems to be common and effective practice for a member of the managing board to take responsibility for maintaining an over-watch. In the largest company, the board might establish a dedicated risk analysis and management department. However it is done, such monitoring is vital, not least to look after the emerging areas of importance in risk terms – social and environmental issues.

Risk as a result of social and environmental issues. There is growing evidence of a new form of risk from social and environmental trends and legislation to supply chains. Certainly in recent years, most risk has made itself felt in purely economic terms – oversight has prejudiced the performance of the enterprise and restored efficiency has been sufficient to relieve this. Increasingly today efficiency alone is not sufficient to restore equilibrium: efficient solutions must take social and environmental concerns into account. For example, climate change was once a matter that needed personal judgment. It remains a controversial issue, but the weight of public opinion has swung towards an acceptance that environmental sustainability is needed in future. This must lead towards legislation and a fresh emphasis on practices that are both efficient and environmentally sustainable. Similar parallels can be drawn over social issues – for example, today it is not acceptable to partner with an enterprise overseas which employs child labour.

All the signs point to rapid developments in the sustainability field, in particular in the immediate future. This can be seen very clearly in the *bestLog* cases – for example, BT's Transport Exchange Group and Sieber's telematics relate directly to the pressures of sustainability issues. Carbon emission measurement, carbon taxation and greater regulation generally are topics driven by considerable social forces and deserve special attention in risk assessment. Many find it helpful to include such issues in their risk registers and there is a strong case for doing so as a means of maintaining a focus on an area of change. Put in a different way, both areas carry with them the implication of greater costs from a variety of sources – through, say, carbon tax, or as operating costs meeting legislation or new standards.

Case:
Sieber – Telematics at SMEs
Sieber decided to use progressive technologies for their truck fleet. So together with a professional provider of telematics systems they developed knowledge in this specific field and can now save on fuel and enhance customer satisfaction by further IT-functions. Additionally they enhanced their employees' motivation as well as level of education

(See Case Collection in Part IV of this book)

Implementation: plans into practice. Earlier, we said that risk management systems are only as good as to the extent to which they are implemented. Do not under-estimate the firmness of purpose that is needed to implement risk mitigation, and the extent of education and training that is involved even in the smallest company. The important factors are simple enough but worth repeating:

- *A clear plan*, well understood at every level.
- *Leadership* towards a risk aware culture.
- *Clear responsibilities* aligned with normal management.
- *Enforcement.* A process of enforcement, making it clear what is to be done and checking and rechecking that it is done.
- *Understanding.* A positive understanding of the contribution that innovation can make to reduce risk and
- *Keep at it.* A commitment to continuity.

This latter factor deserves some attention in its own right, for without continuity you will not get much benefit.

- *Keeping risk reduction alive.* For most of us, keeping things rolling is just as difficult as getting them going. But when it comes to risk mitigation, continuity is more demanding than in many other areas:

- *Early enthusiasm*. In the early stages there are a number of easy wins, but building the knowledge of risk quickly becomes a slog with results coming slowly and piece by piece; and
- *Endurance*. Taking the longer view, there is a risk of boredom and people getting stale. This is the last thing you want for obvious reasons – the bored investigator is certain to miss a vital clue;

The art lies in developing a balance between robust processes that make the routine of risk analysis accessible, but which stimulate interest because they are easy to do, and of clear benefit to the business, preferably not only in terms of risk mitigation alone. Your people will be much better motivated by a process that supports business improvement generally, rather than just supports some seemingly remote risk register.

This is going to need some ingenuity from the outset:

- *Commit to continuity*: It is important to factor in continuity from the earliest stages of getting organised. This is a matter of keeping it in the back of your mind available to contribute to taking a sound decision on doing things a certain way.
- *Communication*: Build the need for continuity in to communication planning – put simply, as results come in, make sure the right people know about them – your people, Tier 1 customers and suppliers, whoever may be appropriate and
- *Leadership*: Make sure that leadership continues to focus upon risk and is seen to do so – the value of example is invaluable here to maintain focus.

This textbook is all about best practice. Having looked at the issues in some detail from principles to implementation we should bring our thoughts together by looking at what best practice in the field of risk reduction looks like.

6.7 Solutions: What Makes the Best Practice in Risk Avoidance?

Earlier in this chapter we made the point that there is no template solution to analyzing and mitigating risk. That said, there are some best practice principles that underpin a successful approach to the matter. They are unsurprising commonsense, but bear repetition precisely because this is the case:

- *Relevance*: Best practice means ensuring that every risk, every process is tested against its relevance to your business – its objectives, position in the market and its resources.
- *Simplicity*: The best plans are simple – they are easier to understand, have a better chance of being adopted by everybody, and are easier to correct when things go wrong.
- *Knowledge*: The most successful know their industry well, and know their particular business in it well. They have a good knowledge of current supply chain practice, they maintain an interest in innovation and the world at large,

and they have a strong professional network amongst those they do business with.

- *Understanding others' points of view*: With risk mitigation, as with many other things, it helps greatly to understand how others see the world and your business. Some trends may be unwelcome, but they have to be faced to prevent them becoming a risk.
- *Analysis*: The best analyse their problem in context before undertaking a particular route to mitigating risk.
- *The best robust framework*: The best put in place a framework that fits their enterprise, is relevant and can be resourced both financially and with the right people.
- *Leadership*: Best practice involves strong and consistent leadership from management, from the top down. The winners inform and inspire which both sustains their risk mitigation process and ensures that the informed risk aware culture they stimulate generates a more effective approach widely applied. This is not just a matter of direction – it is as much about engaging your people in understanding the issues, and listening to them too.
- *Get your own house in order*: Best practice means getting to grips with your practice in the business and working on the risk inherent in how you do things. Those who examine this ruthlessly and regularly are stronger as a result. They do business better and reduce risks by doing so. They place themselves in a much better position to deal with others, not only because of what they know, but also by reason of their better knowledge of what they need to know about those with whom they are in business.
- *Cooperation*: Best practice invariably involves cooperation with others outside the business – initially Tier 1 suppliers and customers, later some of their Tier 1 customers and suppliers – to produce better knowledge of risk and better means to mitigate it and
- *Continuity*: best practice also means building for sustainability by recognising that risk are constantly evolving – one kind of risk goes away, another takes its place – and that risk analysis is not a one time activity. It involves an open-minded and alert review process that roles forward in step with the needs of the business.

The risk management plan. All this adds up to a plan to assess and manage risk in your enterprise in the context of its supply chain. This table will help you to assess whether your plan meets that objective (Fig. 6.7):

6.8 Managing Risk: The Principal Lessons from the BestLog Project

The genesis of the bestLog project lies in the value of best practice to increase efficiency so as to disconnect economic growth from its consequences in the past – such as the inexorable rise in road traffic volumes and the increased congestion

DOES YOU RISK MANAGEMENT PLAN MATCH THIS?	CHECK BOX
•Monitors our business. A regular review of our conduct of business in house,measuring performance in an understandable way,taking in risk assessment as part of its activity, and matching it to best practice.	☐
•Is Relevant. The response must be proportionate, tuned and matched to our business so that it is relevant and no more of a burden than it should be.	☐
•Is part of our business every day. The best practice builds in risk assessment as a part of normal business.	☐
•Is firmly in the minds of leadership. Whilst it essential to make risk assessment part of business as a whole, it must not get obscured by other aspects of business either. Risks come and go in priority,and top management, even in the smallest enterprise, must be aware of changing priorities by being engaged in regular risk review.	☐
•Is everybody's business. Growing a risk aware culture, together with accessible understanding through a relevant risk register is fundamental is vital to engage your people in a proportionate approach to risk in your business and the value of doing things well as a means of mitigating risk.	☐
•Encourages cooperation. There has to be a culture of cooperation, at least in matters of risk, if risk from outside the business is to be controlled. Cooperation of an appropriate kind with Tier 1 customers and suppliers has to be central to planning, implementing and operating a risk control system.	☐
•Is robust. Whatever process is adopted, it must be easy to understand and emplaced in such away that it is not critically dependent upon one individual or vulnerable to failure through circumstances, such as a surge in business activity. By definition, it needs to be tested.	☐
•Is simple.This is not a simple subject, so any process will involve some complexity. But this must not mean that the process needs to be arcane or defined by detail. Rather the origins of simplicity start from a clear and well articulated concept from the outset that is derived from a commonsense relevant approach that suits the particular business in question. In practice, such simplicity contributes hugely to achieving the goals that are already set out.	☐

Fig. 6.7 Assessing your risk management plan. Do you match up?

which results. Insofar as managing risk is concerned, the project has provided an important opportunity to re-evaluate old lessons, and to identify some new ones.

Our work in the project has identified yet again that not a lot is new in this field. The nature of risk comes and goes, but generally it is a matter of emphasis on aspects of risk that are already known: for example, trading conditions may deteriorate and increase risk as a result, whilst risk from fuel price rises may reduce in the same timeframe.

The principal remedies still lie as they always have done:

- Get risk in perspective – the principle of "risk aware not risk averse".
- Get your business processes, both inside and outside your business, as efficient as it is possible for you to do.
- Be agile.
- Be persistent in your vigilance to identify risks and have a plan for dealing with them.
- Engage all your people in the process.

But, at the same time, the bestLog Project has identified some things that are new, and which merit greater priority:

- It is important to factor in sustainability in your work to reduce risk.
- It is no longer possible to ignore environmental and social factors when addressing risk.

- Very close attention to cooperation with others is of increasing importance as new technologies spread through the supply chain. Arguably it is the key to the vigilance that is even more necessary to identify and deal with risk. There are enormous advantages in sharing the burden, and, indeed, any effort to mitigated risk outside the confines of your business is virtually impossible without it.
- Early warning is the key in a world that is developing and changing fast – to make no provision to monitor the world around your business is to court disaster.

Although none of the bestLog cases are concerned solely with reducing risk in the purest sense, they do confirm the view that best practice is one of the principal tools available to deal with risk, and perhaps, even turn it to advantage when others are not so prudent. Be in no doubt that the application of best practice in supply chains is an essential way of reducing the risk to your business.

6.9 Summary

All enterprises are vulnerable to *risks*, both calculated and unforeseen. *Supply chains* and supply chain based enterprises are particularly vulnerable to risk because they must respond to priorities and demands set by their customers, and because the added complexity of the interdependence between the enterprises involved.

Risks to successful business fall into four broad categories: controlled risks of a calculated nature directly as a result of business decisions and which are not considered further here; uncontrollable risk outside our control such as natural disasters; and, two kinds of controllable risk, those arising inside the business, and those coming from customers and suppliers outside the business.

Evidence points to an increase in the risks faced by supply chain businesses overall today. Effective risk management and mitigation is a key foundation for survival in difficult circumstances.

Principles are examined: a balanced and proportionate approach to risk; the need to be risk alert, not risk averse; and, the role of leadership, competence and cooperation.

Against this backcloth, key ways of mitigating risk are reviewed. The importance of getting organised to deal with risk is emphasised. The matter is best approached in a step by step way: first, by examining the risks inside the business, from say, not doing things well; second, by looking at the external risks arising from your Tier 1 or immediate customers and suppliers, extending ultimately to their Tier 1 contacts as well; and, third by looking at the broader uncontrollable risks. These are not necessarily distinct phases: in practice some information from each step will inform the others as you build a matrix of risk information. At the heart of all this is a risk register that maintains an intelligible record of risks, a live document to which all refer when making decisions that are central to the enterprise.

The chapter concludes by concentrating on putting principles into practice – matching your business needs, being properly skilled to make judgements, the cooperation necessary to handle risk reduction outside your business, and ways of dealing with uncontrolled risk. Finally, there is some guidance on what is best practice in managing risk, focusing on a number of commonsense considerations – the necessary relevance of all measures to the business; a sound, well educated basis of knowledge as a basis for analyse; and, a robust simple risk management framework. A checklist is provided at the end.

Appendix: Example Risk Register in Use by Medium Sized Company Commercial in Confidence. Example and Co Risk Register File Reference: Register 20 Effective Date 1 November 2008

Serial	Risk	Risk details	Occurrence[a]		Risk controls	Monitoring, tracking and reporting state
			How likely?	What impact?		
(a)	(b)	(c)	(d)	(e)	(f)	(g)
A.01	Warehouse and Office Buildings	Building protection Trained staff – by law must have 3 fire wardens, 5 First Aid assistants	2	4	InsuranceAnnual Safety Inspection Quarterly Fire practice	Policy renewal: 1 January 2009 1 September 2009* Overdue – chased 151008 OK 010908. Next 011208
A.02	Power failure for more than 1 hour	Lack of power affects IT system and warehouse operations	1	4	Stand-by generator Emergency plan	Tested monthly OK 31108 Tested quarterly – due 011208

[a]Expressed on a scale of 1 to 5 – 5 = high, 1 = low

Tasks:

1) Examine the role of a Risk Register in your chosen company as a means of managing and mitigating risk.
2) Discuss the impact of sustainability issues on risk mitigation over the next 10 years.
3) Road safety continues to be an issue in Europe. Examine how an SME might embrace a road safety aware culture into more efficient supply chain operations.
4) All logistics enterprises depend critically upon information technology, yet many take surprising risks, many of an elementary and avoidable kind. Examine the risks in IT for your chosen organisation and identify possible solutions.
5) Cooperation is fundamental if there is to be any worthwhile relationship to reduce risk with a Tier 1 supplier or customer. On the basis of your chosen enterprise, develop a plan to initiate such a dialogue with a notional Tier 1

client. Assess not only the tasks to be done, but include how, and the proposed timetable to achieve success.

Bibliography

Reading List and References

Christopher M. (1998): Logistics and Supply Chain Management: Strategies for Reducing Cost and Improving Service.

George, M.L., McGraw-Hill (2002): Lean Six Sigma. Combining Six Sigma Quality with Lean Speed.

Peck, H. (2003): Creating Resilient Supply Chains: A Practical Guide. Report produced by the Centre for Logistic and Supply Chain Management, Cranfield School of Management, Research Funded by the UK Department of Transport.

Peck, H. (2003): Understanding Supply Chain Risk: A self Assessment Workbook. Prepared by LCP Consulting in conjunction with the Centre for Logistic and Supply Chain Management, Cranfield School of Management, supported by the UK Department of Transport.

Piotrowicz, W. (2008): Designing Closed-loop Supply Chain: Implementation of the WEEE directive in the UK, The European Retail Digest 56:12–15.

Supply Chain Operations Reference – model (SCOR) Supply Chain Council Version 9 http://www.supply-chain.org.

Further Reading

Business Continuity and Supply Chain Management (2008): Chartered Management Institute of Management (2002) followed up by annual surveys, the latest being March 2008 URL: [http://www.managers.org.uk].

Supply Chain Resilience (2003): Final Report produced by the Centre for Logistic and Supply Chain Management, Cranfield School of Management, for the UK Department of Transport.

Supply Chain Vulnerabilit (2002): Final Report produced by the Centre for Logistic and Supply Chain Management, Cranfield School of Management, for the UK Department of Transport.

Part III
Section C

Chapter 7
Future Sustainable Supply Chains

Richard Cuthbertson

7.1 Understanding the Future

A combination of primary and secondary research was conducted in order to realise the objectives of this project. Secondary research was conducted in the first phase of this project to identify the key issues pertinent to the development of sustainable supply chains. Secondary research informed the next stage of primary research and enabled the researchers to design an appropriate tool for the collection of case studies via key respondent interviews. According to Jankowicz (1995), this technique is especially useful in defining the essential characteristics of some issue by drawing on the personal experience and understanding of the people involved. The interviewees comprised a diverse group of practitioners, operating in many different contexts. The data collected from the secondary and primary research were then analysed to develop the various scenarios of future sustainable supply chains.

Scenario planning is a strategic management process used in the private, public and charitable sectors, as documented by Van der Heijden (2009). Scenarios are plausible yet challenging stories about the future that address a core issue of importance, such as the future of supply chain management, for a particular set of stakeholders, including practitioners in a wide variety of circumstances and policy makers in different national contexts. By giving these diverse stakeholders a shared basis for discussion, scenarios enable creative thinking about how to shape the future of sustainable supply chain management. Scenarios also encourage the discussion of uncertain events. By making uncertainty explicit, preconceptions may be identified and help decision makers consider how they and others might react to the different ways in which the future may develop. Once described, scenarios can be used for continuing discussions about issues such as risk management and contingency planning.

Scenario building is primarily a qualitative activity, focused on generating challenging insights. The scenario process has eight steps as follows (Fig. 7.1):

C. Tyssen et al., *Sustainable Supply Chain Management*,
DOI 10.1007/978-3-642-12023-7_7, © Springer-Verlag Berlin Heidelberg 2011

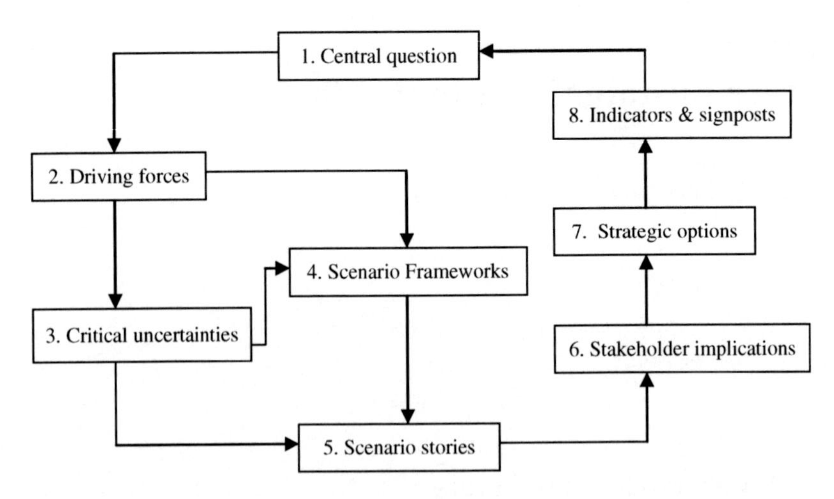

Fig. 7.1 The eight step scenario approach

7.2 The Central Question for Sustainable Supply Chains

The key questions for the future development of sustainable supply chains arise from the potential different scenarios that may develop due to the differing combinations of corporate decisions, both inside and outside the firm, as well as the role of public policy and the various standards bodies and third party influencers.

7.3 The Driving Forces that Will Influence Sustainable Supply Chains

The driving forces behind current supply chain development are identified by mapping the strategic environment of the supply chain sector. This is a twofold exercise and entails both mapping the past and looking forward. This is largely discussed in the first chapter of this book.

The manufacturing of products has become more centralised, usually based on either low cost countries or high investment technologies. This has resulted in a reduction of the total number of factories, and a move away from nationally-based production towards single locations manufacturing a particular product for the whole of a region, or even the world. This has increased the demand for freight transport. By rationalising their supply base, companies lower their overall production cost at the expense of increasing the logistics and transport services within the supply chain. This concentration of suppliers is particularly utilised in large scale industries, where the producer may have a large degree of power within the supply chain, for example in the automotive industry. Similarly, this has resulted in a reduced number of stockholding warehouses. These can yield a large financial benefit, outweighing the cost to the firm of any additional transport required, but has consequences for the environment and society.

At the same time, markets for products are extending, particularly as some of the world's largest potential markets, such as India and China, begin to develop rapidly in terms of consumer demand. The combination of the wider geographical sourcing of supplies and the wider distribution of finished products is extending supply chains, both upstream and downstream. These twin driving forces of centralised supply and diversified demand have been supported by the information technology and systems that integrates these extended supply chains in real-time. This trend towards global rather than local supply chains is also a result of companies focusing on their core competencies, and outsourcing non-core activities.

Reverse logistics has been helping to alleviate some of the environmental and societal concerns, for example through more efficient backhauling and greater recycling capability. However, reverse logistics may also increase transport demand and potentially reduce utilisation rates. Similarly, while technology has reduced the need for transporting regular mail due to the development of electronic mail services, the demand for small units, such as parcels, has increased due to online retailing activity and customer-to-customer transaction websites, such as EBay – thus potentially increasing freight transport requirements.

Supply chains have had to become more responsive and more flexible enable firms to compete in this global market. This, in turn, has impacted the freight modal split, with the result that road transport (the most flexible form of transport) has continued to take an increasing share of the expanding demand for freight transport. Changing cost structures, better infrastructure, the standardisation of loading units and deregulation have all played their part in supporting this driving force.

Thus, as demand for freight transport has grown, the impact on society and the environment has been questioned, along with the role of legislation, standardisation and financial motivations to move towards more sustainable supply chains.

Based on the results of the case studies and the secondary research, these key driving forces were identified. It is important to mention that these driving forces are neither exhaustive nor mutually exclusive. Their role is to inform the scenario-building process by identifying the most important and most uncertain factors, in the belief that how they might interact will help stakeholders build a more comprehensive picture of possible future worlds.

7.4 Considering the Critical Uncertainties in Supply Chain Management

Critical uncertainties are those driving forces that are both highly important and highly uncertain. Numerous critical uncertainties surround the future development of sustainable supply chain management. Identifying the critical uncertainties enables us to select scenario dimensions that are the most critical driving forces.

As can be seen in the previous chapters, to pursue a sustainable supply chain strategy four key areas and their associated risk must be considered: customer demand, performance measurement, people development, as well as standards and infrastructure constraints.

7.4.1 Customer Demand and Supply Chain Strategy

While there is some customer awareness around sustainable supply chains, particularly "green", environmental issues, and, to a lesser extent, social issues, there is always a clear focus on the supply chain as an economic cost, as reflected in the price of supply chain services or as part of product price. Therefore, going forward, there is scope to improve sustainable awareness, but there is also the threat that this may lead to excessive pricing differentials between those firms focused solely on short-term costs and those firms with a longer term view of sustainable supply chains.

As shown in the Chap. 2, these long-term trends pose challenges for supply chain managers. The combination of global supply and demand creates requirements in terms of security, environmental protection and the utilisation of limited resources. More than ever before, modern supply chain managers are confronted with dynamic, complex and inter-related global supply chains increasing the strategic importance of management decision-making. In order to respond to the environmental and social challenges and yet remain competitive, supply chain managers need to be able to identify and understand new developments and their potential impact on their company, customers as well as the wider business environment.

7.4.2 Performance Measurement and Operations Management

Similarly, while there is much greater access to advanced technology, information and communication systems, performance measurement systems are still largely local and focus on economic measures, rather than social or environmental measures. There is significant scope for improvement here but the cost of introducing new and integrated systems and technology is expensive and often prohibitive for many within a supply chain network. Hence, there is a great deal of uncertainty around this area.

As we saw in Chap. 3, performance measurement and management are necessary to monitor how a selected strategy is realised and goals achieved. Again, it is important to stress that performance is not only about the financial outcome generated by the company, but also about the performance of the whole supply chain, including social and environmental measures, as well as the traditional economic-based measures. Performance measurement is necessary to manage performance, to take actions based on data which reflects company performance and its external impact on sustainability. Collected information about past and current performance should be communicated to all relevant parties and used to influence future decision making. Any measures collected need to be reviewed on a regular basis and changed if necessary.

Measurement is a necessary activity to manage and improve performance across the whole supply chain, even though, in reality today, it is difficult and unusual to cover all supply chain members.

7.4.3 People Development and Managing the Organisation

As with technology, it is less the lack of capability than the lack of investment available that affects the pace of people development. There is no doubt that well-educated and talented individuals exist, however high labour costs in developed parts of the world, such as Europe, combined with a cost focus in supply chain management constrains the overall level of talent and level of education. In particular, this combination creates uncertainty in the level of education and capability of supply chain managers to fully appreciate the increasingly long and complex supply networks within which they need to operate. As identified in Chap. 5, the modern supply chain manager does not just manage logistics processes, but also people, products, knowledge and infrastructure. Perhaps more importantly, such issues must be managed across the supply chain, within and between companies, for the goal of sustainability to be achieved. This requires consideration of cross-functional and cross-company activities, requiring many general management skills and knowledge, as well as the more specialist supply chain elements. So, successful supply chain managers are required to integrate issues of sustainability into their day to day business decisions, as they manage (i.e. design, plan, organise, implement, supervise and control) processes, products, knowledge, infrastructure, and people, while accounting for the overall balance of social, economic, and environmental impacts. Hence, as a direct consequence of their general and pervasive character, successfully managing supply chains in a sustainable manner requires the identification, evaluation, and prioritisation of many different issues, each unique for a company at any given point in time.

While this challenge is immense, any uncertainty around people development is lessened by the fact that these are primarily elements that are directly under the control of any firm operating within the supply chain, for example through effective recruitment and subsequent development policies. As in any business, the people are a major factor in driving success, whether the goal is sustainability or not. Any successful manager has to motivate employees to support necessary changes, and the same is true when focusing on sustainable supply chain management. Likewise top management has to understand the measures taken and support them. Here it may be especially important to argue the long-term nature of sustainability actions, as a positive impact on financial results may not be directly visible in the short term. The commitment of top management is crucial for any investment required to pursue sustainability goals, especially as the revenues may never be directly observable.

7.4.4 *Standards and Infrastructure Constraints Imposed Outside the Organisation*

Unlike people development, uncertainties around standards and infrastructure constraints are not within the direct control of the organisation, though they may be influenced by them through industry lobbying and so on. Infrastructure development may act as a constraint on all competitors equally, while legally enforceable standards may constrain the poor performers from operating when such standards are relevant but may constrain the good performers when they are not relevant. Thus, there is a high degree of uncertainty around such issues. Hence, there are always opportunities to improve on existing standards and these may prove hugely important in the drive towards greater sustainability across all supply chains, whether legal or voluntary. However, this greatly depends upon the relevancy of such standards. Irrelevant standards are a threat to progress towards sustainable supply chains. This is a particular threat in cases where the standards are not easily visible or exposed, for example non-legal standards imposed upon a supply chain by a dominant player that may make good sense for the dominant player but not for the overall supply chain. This example is a very real concern for small and medium sized enterprises, who generally cannot afford the same degree of investment in time, money and other resources that large businesses may incur.

Thus, these issues of standards and infrastructure development over which there is little direct control, and often little influence, require firms to be aware of and anticipate the attitudes and actions of other relevant actors, including suppliers, service providers, customers, governments, local communities, and NGOs. As shown in Chap. 6, managing for such externalities is an extremely complex process. The regulatory environment of supply chains has become highly complex and increasingly challenging for supply chain managers. The best practitoners regularly modify and adapt the management and design of their supply chains to reflect these changing conditions.

Bibliography

Van der Heijden, K. (2009) Scenarios: the art of strategic conversation, John Wiley & Sons. 2nd edition.

Part IV
Section D

Chapter 8
Section Structure

Part IV focuses on the cases for illustration. It is dedicated to further ellaborate and examplify the points discussed in Part II. Cases are drawn from real-life business situations and were raised during the project duration. The cases are meant to demonstrate and illustrate cases of good practice, where the three dimensions of sustainability where fulfilled and hence improved social, ecological and economic operating figures of a company.

In order to assort the cases in a sensible structure we used a structure provided by one of bestLog's workpackages. There we developed a pyramid which will in our context work as the *BestLog* bestLog Case Navigator.

The case navigator tackles all relevant areas to supply chain management. The cases are hence structured accordingly (Fig. 8.1).

Derived from the "Business decisions"-model of Alan McKinnon the bestLog pyramid distinguishes mainly on the planning time horizon of decisions.

- Strategic decisions (number, location of production/ distribution facilities)
- Commercial decisions (suppliers, customers and sub-contractors)
- Operational decisions (scheduling production and distribution)
- Tactical decisions (day-to-day management of logistics function)

The four levels from the hierarchy influence each other sequentially. Decisions on one level set the framework for all decisions on the subsequent levels. Therefore decisions made on the strategy level influence all other steps of the pyramid whereas decisions on the support level not necessarily influence the logistics planning decisions.

All dimensions are surrounded by different stakeholders as suppliers, customers, logistics service providers and last but not least the public.

The section for strategic decisions focuses in long term changes in a company's orientation and future. The according cases contain long term decision that are eventually cost intensive and amortize on a long term view.

Logistics planning cases tackle less long-term focus but still contain major changes in the company's concept. They cover changes in network design or tactical scheduling.

C. Tyssen et al., *Sustainable Supply Chain Management*,
DOI 10.1007/978-3-642-12023-7_8,

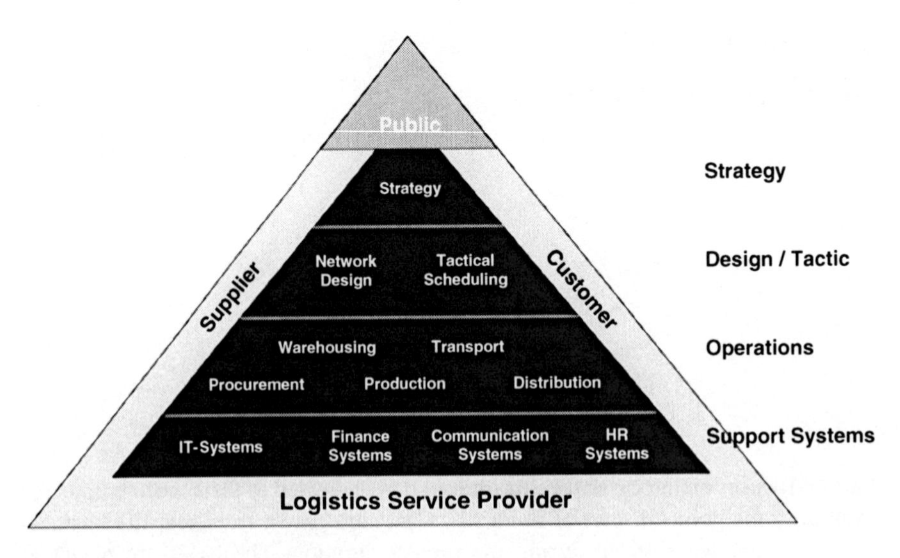

Fig. 8.1 The bestLog case navigator

Logistics operations cases deal with the optimization of warehousing or transport processes as well as changes in procurement and production or distribution. Companies that for example used a new production system resulting in less transports will be put here.

The basis and lowest layer of the bestLog Case Navigator covers support topics. These topics are for example enhancements in IT- or communication-systems, reflecting a rather technical area, whereas financial and human resource systems describe rather commercial fields.

Readers can either select cases by the references used in the text of Part II or, in order to tackle specific topics, choose a managerial dimension described in the bestLog Case Navigator.

Chapter 9
Markets and Strategies Cases

9.1 Strategic Environmental Plan: INDITEX – Pro-Kyoto Project

Sonia Guerola Pérez

INDITEX is developing a Strategic Environmental Plan 2007–2010 which is divided into five specific projects:

- INDITEX Pro-Kyoto Project
- Sustainable Store Project
- Energy and Environmental Integration Project
- The Carbon Footprint Project
- The Terra Project

We will focus in the Pro-Kyoto Project later in this case study because the sustainability of logistics operations is one of INDITEX's key business values. The group's commitment to sustainability, climate change mitigation and the Kyoto Agreement is specifically focused on the reduction of greenhouse gases generated by transportation.

9.1.1 Problem Description

INDITEX is one of the world largest fashion distributors, with eight brands – Zara, Pull and Bear, Massimo Dutti, Bershka, Stradivarius, Oysho, Zara Home and Kiddy's Class.

The Spanish government and European Commission have increased fines for excess CO_2 emissions. INDITEX's own CO_2 emissions were considerable. INDITEX also realises the importance of brand image to consumers, especially

C. Tyssen et al., *Sustainable Supply Chain Management*,
DOI 10.1007/978-3-642-12023-7_9, © Springer-Verlag Berlin Heidelberg 2011

nowadays in terms of sustainability and environmental responsibility. Moreover the Spanish government is funding companies' efforts to become more sustainable.

With these factors in mind, INDITEX is developing a Strategic Environmental Plan 2007–2010, which is divided into five specific projects: the Sustainable Store Project, INDITEX Pro-Kyoto Project, Energy and Environmental Integration Project, The Carbon Footprint Project and The Terra Project.

The Pro-Kyoto Project is focused on logistics operations, whose sustainability is one of INDITEX's pre-eminent business values. The Group's commitment to sustainability, climatic change and the Kyoto agreement is specifically focused on a targeted 20% reduction of greenhouse gases generated by transportation.

9.1.2 The Solution

The main goal for the Pro-Kyoto project is to reduce transportation greenhouse gas emissions. To achieve this goal, INDITEX has established four initiatives that have already been implemented, and will be developed until 2010:

- A Bio-Diesel Programme: Zara's entire fleet of transportation vehicles will run on bio-diesel, for which a supply network will be set up at loading points and along trucking routes. The vehicles concerned are operated by third party logistics contractors.
- Fleet driver training courses on fuel-efficient driving form part of a plan implemented in conjunction with logistic suppliers. It also stipulates that all vehicles in the Zara fleet must comply with the European Commission's EURO 5 NOx vehicle emissions standard two years ahead of EURO 5 coming into effect in 2009.
- Zero-emissions electric vehicles will be used in factories and logistics centres to replace an older petrol-fuelled fleet.

The second project is the Sustainable Store project. It has involved developing criteria for sustainability and the efficient use of energy in retail outlets, through three main strategies:

- Eco-efficient store model: INDITEX, in conjunction with the University of Vigo, is designing an ecological energy management model for retail outlets. Lighting equipment will be changed to equipment that offers high efficiency and low energy consumption.
- Sustainability of packaging and wrapping materials: Work in this area includes eco-certification for all plastic and paper bags used by the Group, (using international standards such as FSC or PEFC) and the use of biodegradable materials in all plastic bags. It also covers all store waste management, which is carried out by authorized agents, and the re-use and recycling of materials such as cardboard boxes, hangers and computer-generated waste.
- Training for all store personnel to raise awareness of environmental impacts.

The third objective is to integrate Environmental and Energy actions in the production facilities. The aim of the Energy and Environmental Integration Project is to introduce renewable energy sources, such as wind, solar and solar-photovoltaic energy in the Group's logistics facilities.

INDITEX's facilities at Arteixo (A Coruña) house the headquarters of the Group itself as well as those of the Zara, Zara Home and Kiddy's Class brand operations, the Group logistics centre and 11 textile factories.

Inditex has implanted a new integrated energy system at Arteixo that includes a 5,000 kW cogeneration plant, a 1,500 sq ft solar installation and a 850 kW wind-turbine. This system covers more than 50% of the Arteixo site's energy requirements.

The Group aims to install similar energy systems in its other main facilities. The plan involves implementing the following actions:

- Solar photovoltaic installations in the logistics centres located at Meco (Madrid), Zaragoza and León, and in the Group's headquarters in Tordera (Barcelona) and Elche (Alicante).
- Minimise carbon monoxide emissions by replacing oil with natural gas for heating and for the production of steam for ironing equipment at Arteixo and Sallent (Barcelona).
- A new cogeneration plant in Tordera.

The Carbon Footprint Project objective is the fourth element of the plan; Inditex will measure the carbon footprint of each of their manufacturing processes and adopt strategies to reduce it.

The Terra Project is the final element in the Strategic Environmental Plan 2007–2010. INDITEX considers tree planting as one effective way of closing the energy/CO_2 loop. The objective is to create the forest mass needed to absorb 100% of the emissions generated by the consumption of electricity at the central office in Arteixo. (A site for this forestation project has yet to be selected).

INDITEX will achieve a reduction in greenhouse gas emissions and increase the proportion of its energy demand supplied by renewable energy sources. It will maximize the efficient use of energy in manufacturing, train staff in environmental management issues and raise their awareness of sustainability.

INDITEX considers it should integrate a set of sustainable development criteria in its working practices, to ensure effective, responsible resource management and the protection of the environment to satisfy society's requirements.

9.1.3 Challenges

Challenge 1: To achieve a 20% reduction of CO_2 emissions. This is an important objective, because to achieve this the Group had to initiate change in several activities, including some of its logistic strategies.

Challenge 2: No commercial availability of biofuels in Spain.

Challenge 3: Scepticism of personnel and fear of technology change (vehicles running on biofuels and electrical vehicles).

9.1.4 Lessons Learnt and Success Factors

- Lesson 1: Logistics and environmental efficiency go together hand in hand. Thus if they are both managed in an integrated way, economic, social and environmental benefits will result simultaneously.
- Lesson 2: Sustainable corporate investment decisions on technological and productivity improvements must value and account for environmental variables.
- Lesson 3: When staff are fully involved in projects of this type, success is more easily won.
- Success factor 1: INDITEX has aligned its logistics strategy with sustainability criteria to realize environmental improvements which can enhance the image of the company communicated by media.
- Success factor 2: Truck drivers have been trained in efficient driving techniques and other employees have taken part in training to make them aware of the new strategy and to encourage their participation.

9.1.5 Benefits

9.1.5.1 Economic

Perceptions of the Group's service and its brands will be enhanced by the company's environmental good practice. This practice prioritises energy efficiency in plants, warehouses and trucks but without negative implications for production volume, productivity; reduced energy consumption produces increased economic efficiency

9.1.5.2 Environmental

Group-wide CO_2 emissions will be reduced by 850 tonnes per year due to the use of biofuels (accounting for 80% of the reduction), and to the use of renewable energy sources and tree planting. Emissions of CO, sulphur and nitrous oxides are also reduced by switching trucks from diesel to biofuels. Moreover, more crops are needed to produce biofuels and to plant trees for the Terra project; this has impact on land use. With the Ateixo cogeneration plant, more than 80% of the energy in the plant's fuel is converted into usable energy. Cardboard production and office waste has been reduced packaging policies will give preference to the use of recycled and biodegradable products.

9.1.5.3 Social

The Group's truck drivers will improve their driving techniques to enhance safety, reduce emissions, while staff in the Group's offices and retail outlets have learned to work in more sustainable ways. These staff development measures have an impact on employment, since it has become necessary to hire new personnel to develop this ambitious project.

9.1.6 Transferability

9.1.6.1 Transferables

Large companies such as INDITEX can afford to invest in this kind of project to protect the environment and yield profits. The renewable energy chosen depends on the geography (if there is profitable wind, sun, water, etc.) but if it is chosen carefully, at least one renewable energy source can be used.

9.1.6.2 Limitations

A project such as INDITEX's Strategic Environmental Plan 2007–2010 can be implemented in any country and in any sector, taking into account the appropriate source of renewable energy in each case.

Chapter 10
Structure and Planning Cases

10.1 Consolidation and Consignment: Sieber – Consolidation of Transports

Christoph Tyssen

The logistics service provider Sieber is well located in terms of transport connections. The company offers warehouse capacity and expertise in warehousing management and consolidating deliveries for its customer. Consolidated consignments of production components are shipped to the customer when needed in the required volumes and combinations. This service avoids any need for additional warehousing capacity at the customer's site and reduces the number of deliveries to the customer's relatively inaccessible production site.

10.1.1 Problem Description

Wishing to concentrate on its core competencies, Sieber's customer, a medium-sized electronics manufacturer, did not want to invest in new warehouse facilities, but rather to invest in its production capacity. The customer wanted elements of its procurement processes to be executed by an external service provider in order to reduce the complexity of its own procurement activities.

The production site of Sieber's customer is in the Swiss mountains, in an unfavourable location in terms of transport efficiency. A rising volume of deliveries to this location meant not only increasing logistics costs, but also added risk to driver safety, as the access roads were unsuited to dense traffic. These conditions led to Sieber and its customer collaborating in the search for a solution to this problem. Sieber itself operated a warehouse in Switzerland located near the Austrian and German borders, in a more attractive location for transport and logistics purposes than the client's site.

C. Tyssen et al., *Sustainable Supply Chain Management*,
DOI 10.1007/978-3-642-12023-7_10, © Springer-Verlag Berlin Heidelberg 2011

The Sieber warehouse is easily accessible with good highway connections. In addition, both Sieber and its customer looked for solutions to increase the efficiency of the procurement and transport processes in the customer's supply chain more efficient. New information systems were analysed in relation to their capacity to serve this objective. As both companies were strongly orientated not only towards efficiency improvements but also to addressing sustainability issues, they favoured a solution which would generate fewer deliveries along with rising volumes of procurement.

10.1.2 The Solution

Both partners realized that it was mutually advantageous in terms of efficiency and environmental concerns to use existing, modern warehouse capacity providing better transportation connections.

Sieber was able to offer dedicated warehousing capacity to the customer, which could be adapted flexibly to the volume and dimensions of the goods the customer needed to stock. Sieber also took over the organisation of deliveries, consolidating and delivering product components in the right sequence to the customer. This solution reduced the number of deliveries through the mountains by factor of 100. (What had been about 210 deliveries per week to the customer could now be concentrated in two deliveries from Sieber's warehouse.)

In order to provide real-time, secure data on the customer's procurement processes, an IT system was installed linking the two companies. Sieber's large, modern warehouse offered an advantage in terms of flexible capacity, as it had been built to work for multiple customers. The customer was able to avoid investing in the construction and maintenance of in-house warehouse facilities, While the client's requirements helped Sieber optimize the efficiently utilization of its own warehouse capacity. The Sieber warehouse is built to current building efficiency standards, and is equipped with energy-efficient technology, such as lighting which, for example, turns lights on in a given aisle of the building and turns them off if automatically when it is not in use. This commitment to energy efficiency was a significant factor in persuading the customer to work with Sieber.

The partnership offered several new opportunities. Firstly, Sieber's warehouse was easier and safer to reach than the client's site, and additionally offered the opportunity to consolidate production component deliveries. Only one truck needs to go to the customer's production site twice a week, reducing truck journeys and their environmental impacts and safety risks.

The customer also faces less complexity in coordinating procurement from different suppliers. The customer transmits its orders to Sieber, which arranges the corresponding consignment and supports the associated organisational structures. The customer also benefits from increased flexibility, with no cash tied up in logistics assets, paying only for the warehousing capacity needed at any one time.

Sieber faced a number of challenges in achieving these customer benefits. In particular warehouse capacity had to be analysed with care to ensure the allocation of enough stock room at all times. The volume of goods that had to be stored for the customer had to be known in advance or at least accurately estimated. The size and weight of all the products had to be known in advance. It was important to know which goods required special storage conditions as dangerous goods or which had to be stored at a specified temperature or humidity level.

Connecting with the customer via modern information systems proved crucial for the success of the project, to share information on orders and procurement requirements in real time to avoid delivery shortages.

All these challenges can only be handled if the collaboration is based on mutual understanding and a common desire for improvement. Especially in a case like this, the quality of information handling is crucial, and only realizable through real collaboration.

This example of good practice shows how the cooperative development of solutions using existing structures can be used to improve the logistics and procurement processes of individual companies.

10.1.3 Challenges

Challenge 1: Organisation of transport to the new warehouse and the management of the associated processes

Challenge 2: Handover of procurement – information provided by the customer to Sieber

Challenge 3: Allocation and process design of warehousing capacity at Sieber

Challenge 4: For SME LSPs it is essential to demonstrate the requisite IT and organizational capabilities

10.1.4 Lessons Learnt and Success Factors

- Allocate the required warehouse capacity at an early stage
  - To guarantee a smooth change-over to outsourced logistics, available capacity must be analysed and facilities dedicated at an early stage.
  - The capacity provided to the new customer must to be aligned carefully with existing customers' requirements.
- Know just what kind of products need to be stored.
  - It is crucial to know what volumes and dimensions of products will be stored in the warehouse.
  - Different products may require special treatment and storage conditions.
  - Personnel must be trained to handle specific products competently.
- Provide energy-efficient warehousing to convince the sustainability-oriented customer

- Collaborate with the customer in order that both partners understand each other's needs and develop mutual beneficial solutions
- Create dedicated solutions for each customer
  - Involve the customer in the development of the solution
  - Demonstrate the advantages
  - Offer a unique solution
- Use and know current IT standards.
  - Information processing is crucial
  - The ability to link the customer's procurement processes of the customer with the information system at the warehouse is essential.
  - Adherence to the latest IT standards helps improve data processing.
  - IT know-how demonstrates the standard of the logistics service provider and enhances employees' skill levels.
- As the customer in this case is itself is a medium sized company, and like the supplier, based in Switzerland, close collaboration proved crucial. As an SME, Sieber could understand and respond effectively to the customer's needs.
- A long-term relationship between the logistics service provider and the customer was the basis for this collaborative solution
  - Sharing information on procurement data requires trust between the partners in outsourcing
  - The customer's dependence on the logistics service provider entails intensive coordination through meetings and confidence-building- activities.

10.1.5 Benefits

With this innovative conception Sieber has developed the quality of its service. The Sieber warehouse represents a buffer for the customer's components stock and facilitates improved procurement performance. This results in fewer deliveries and less complexity for the customer's goods inward department. Sieber also takes care of all customs clearance for the customer's supply chain.

10.1.5.1 Economic

In cost terms, the consolidation effected by Sieber significantly reduces the number of deliveries to the customer's production site. Compared to a previous 210 deliveries per week, only one is now required. Furthermore the complexity of procurement processes at the customer's site is reduced, while the incremental volume means that Sieber can operate its existing warehouse with a higher level of efficiency. Each partner concentrates on its core competencies and may thus generate economies of scale. Greater flexibility in stocking helps the customer to quickly adapt production volume to fluctuating market demand.

10.1.5.2 Environmental

The main impact in terms of sustainability was the reduction from a large number of truck deliveries to the customer to just two per week. This reduced the amount of fossil fuel consumed and thereby the emission of pollutants including CO_2. Sieber already managed a warehouse, so the use of land for a new warehouse at the customer's production site could be avoided. Furthermore Sieber already operated a warehouse with a more or less fixed energy consumption. Building a new warehouse at the customer's site would consume a large amount of energy, both in the construction of the new warehouse and in its maintenance.

10.1.5.3 Social

The provision of specialized services requires creative and well-trained employees working for the service provider. Well educated personnel are needed, since any innovative idea has to be developed before it can be offered to a customer. Later on, the customer's own planning and warehousing knowledge needs to be transferred to the logistics service provider. This requires training for the logistics service provider's employees. The long-term contract negotiated in connection with the consolidation solution secures long-term job security for staff. The social dimension also benefits from environmental improvements, inasmuch as noise and air pollution are reduced by fewer truck journeys.

10.1.6 Transferability

10.1.6.1 Transferables

Every SME in transport and logistics could offer the services described in this case study. The SME needs to be able to demonstrate know-how in warehousing and data processing. It needs to be able to allocate spare space to store a new customer's stock. With these provisos, this solution is highly transferable, particularly since similar solutions can be implemented independent of the company's size.

10.1.6.2 Limitations

The company needs know-how as well as warehousing expertise and capacity. There has to be a customer base that require this kind of consolidation and order fulfilment service.

10.2 Supply Network Redesigning: Shell Chemicals Europe and Bertschi AG

Christoph Tyssen

The case of Bertschi AG and Shell illustrates how the two companies successfully redesigned the supply network of Shell's petrochemical plant in Wilton/UK. Prior to the beginning of this vertical collaboration, Shell was facing issues regarding the storage of finished products. Due to the number of specifications for Shell's chemical products and the requirement to store them separately, the company faced fixed costs which were excessive for the company's day-to-day needs. Jointly, Bertschi and Shell developed a new concept, allowing both companies to share the risks while generating individual benefits. By creating a flexible stock system for the chemical products, Bertschi was able to offer effective services, while Shell profited from greater flexibility and shorter response times to customer orders.

10.2.1 Problem Description

Generally, Shell was facing the problem of insufficient storage capacity for the chemical substances produced at its site in Wilton (Northeast England). The increasing number of specifications in the chemicals portfolio and the constantly growing volume of produced goods were the main drivers for the redesign.

Due to the unpredictable order patterns, Shell was forced to hold buffer stocks in off-site facilities, based on medium/long-term contracts. Prior to the solution presented later in this case, 75% of the finished goods produced in Wilton were stored off-site. This created high fixed costs for Shell which were not in sync with the company's day-to-day business.

Another factor adding complexity to the challenges Shell faced was the requirement that the composition of each batch be certified. The many reloadings during a product's journey en route to the final customer and the required certification made it difficult for Shell to respond to short-notice orders. Furthermore, the necessity to clean each tank before filling it with another substance increased lead times for product deliveries as well as the costs for Shell.

In order to overcome those challenges, Shell faced the decision to either expand the storage capacities in Wilton or to implement a new concept in order to solve the aforementioned problem.

10.2.2 The Solution

Basically, Shell had two main options to reduce complexity in their production and storage processes and therefore also reduce the lead times to the final customer.

They could either expand their storage capacity in Wilton or try out a new concept. Due to the high cost and limited flexibility of an increase in storage capacity, Shell decided to create a new concept together with a logistics service provider.

After receiving Shell's request, Bertschi developed a clear approach towards the problems Shell was facing. The proposed concept envisioned the direct loading of the finished goods from the production tanks into tank containers 24/7. In order to make the operational processes more efficient, it was important that Bertschi's concept should fill the chemicals produced by Shell into transportable tank containers and therefore permit effective mobile storage. Eventually, the tank containers are either delivered directly to the customer or stored according to the build-to-stock principle if there is no connection to a current customer order. To enable a quick response to short-notice customer orders, a safety stock of 28 transportable units is always maintained.

Despite this safety stock, the overall inventory was reduced by more than 35%. In order to further accelerate the loading process, Bertschi invested in the training of their drivers. They were trained to self-load their vehicles. This investment not only accelerated the loading process but also made the drivers more aware of problems occurring at the operational interface between the two companies. Furthermore, the self-loading of vehicles presented Bertschi with an opportunity to gain Shell's trust, as their drivers would take on another important function in Shell's operations.

In order to implement its concept, Bertschi had to acquire a new site. And in order to respond quickly to short-notice orders and/or changes in Shell's production plans, this site needed to be close to the Shell site in Wilton. Bertschi therefore decided to acquire a site in Teesside. To ensure smooth operations, this site had to meet all relevant health and safety standards as well as the legal stipulations of course. Since Bertschi had to invest several million Euros, it was essential to have the security of a long-term contract between the company and Shell.

The process aimed at increasing transport efficiency required Shell to standardize the size of its produced batches to 50+-tons. This enabled Bertschi to improve its transport efficiency by loading two 25-ton tanks rather than splitting the orders into 23-ton loads. Despite the benefits of this strategy, Shell took a risk by standardizing the order weight for its customers.

Moreover, Bertschi contributed to increased transport efficiency by improving return loads within its tanker fleet. They collected raw materials from Shell's plant in Stanlow (Northwest England). Whenever possible, these loads were compatible with the outgoing finished goods, thereby avoiding empty runs as well as expensive and time-consuming cleaning of tanks.

To ensure smooth implementation of the concepts described above, an employee from Bertschi was "embedded" within the Shell operation at Wilton while the site in Teeside was still under construction. This enabled Bertschi to gain insights into the operation run by its customer and to collect information on Shell customers and their specific delivery requirements.

The integration of the Bertschi employee also provided an opportunity to apply specific company knowledge rather than utilizing standard solutions. The "embedding" of a Bertschi employee and the insights gained into Shell's operations demonstrate the highly collaborative and trustful relationship. This attitude was maintained throughout the further development of collaboration. For example, the partners chose to implement a communication concept which transcends not only company boundaries but also traditional hierarchical levels of management.

The result is that the teams in both organizations work together on an internal and external level. This ensures that operations are agile and responsive, resulting in the provision of the best possible service to the final customer. The personal interaction between the two companies is based on the "management of exception" concept, and information exchange therefore mainly takes place via an IT interface.

The IT system used by Shell enables the company to manage orders, production forecasts, sales forecasts and customer data.

If an order is received at Shell, the information is transferred directly to Bertschi, where the order data is downloaded on the internal order processing system. Afterwards, the order is tagged with an internal number, which allows the order to be traced throughout Bertschi's operation.

In conclusion, it can be stated that so far Bertschi has been able to convince Shell of the advantages of the new transportation concept. The expansion from one to twelve product specifications handled by Bertschi is a good indicator for this.

10.2.3 Challenges

Challenge 1: Building trust between the players: the most important requirement for the success of collaboration was the continuous building of trust between Shell and Bertschi. This process formed the basis for overcoming the challenges mentioned below.

Challenge 2: Information sharing: the sharing of information such as sales data and production figures was one of the key success factors for cooperation between the two companies. Implementing an integrated IT system was essential in order for Shell and Bertschi to operate their day-to-day business processes efficiently.

Challenge 3: Knowing the customer: prior to providing Shell with a tailor-made solution, the Swiss LSP had to find out about Shell's requirements.

Challenge 4: Measuring the success of collaboration: in order to convince Shell of the benefits of the solution and to identify improvement levers, it was essential to implement the right KPIs in order to measure the success of collaboration. To foster the trust-building process, the KPIs were chosen by Shell.

10.2.4 Lessons Learnt

- It is essential to have the ability to rapidly respond to changes in the customer's production plans

- Moreover, it is crucial to gain and maintain the customer's trust in order to fully implement the business model

10.2.5 Success Factors

- Close cooperation across different levels as well as the creation of joint project teams are essential for the sustainable success of the company
- Especially on the operational level, intensive communication is crucial in order to identify potential difficulties without delay and to overcome these difficulties
- To ensure the success of collaboration, it is important to closely monitor operations, especially at the interface between the partners
- If difficulties become apparent, it is important to not only rely on standard concepts used throughout the industry but also to exploit individual knowledge within the partner companies in order to provide tailor-made solutions.
- Despite the importance of communication within the project teams, the implementation of a compatible IT system is equally important in order to ensure a smooth flow of information between the companies involved in the collaboration process.
- In order to "be on the same page" regarding the developments in the collaboration process, the definition of jointly agreed KPIs and fast reaction to changes in these KPIs is important.
- If the goalposts of the customer shift, the previously defined KPIs should be modified in order to maintain the provision of a tailor-made concept to the customer.

10.2.6 Benefits

The quantifiable benefits of the collaboration are indisputable. Quality control savings exceeded 50%. Inventory has been reduced by 35%, administrative overhead has been reduced by 50%. The venture has driven production capacity improvements, which has allowed better weight utilization in the loads delivered. Additional savings in supply chain costs were achieved through more efficient planning. Furthermore, the improved service level is perceived as an important benefit.

10.2.6.1 Economic

- Due to the decrease in the amount of necessary decanting, quality control savings exceeded 50%.

- The direct filling of finished goods into transportable units led to a reduction in inventory, which has been reduced by 35%.
- Moreover, the number of required certifications was reduced, which led to a decrease in administrative overhead costs of 50%
- The increase in efficiency with regard to the utilization of transportable units led to a reduction of more than 35% in supply chain costs
- This increased efficiency and the ability to respond quicker to short-notice orders has improved the level of service to end-customers
- The simplification of the filling and transportation process made higher plant capacity utilization possible.

10.2.6.2 Environmental

- Preference for intermodal transport, increased amount per load and decreased total number of transports per year.
- Furthermore, greater use of intermodal transport led to a decrease in CO_2 emissions.
- The number of tank cleaning operations was significantly reduced thanks to the new concept.

10.2.6.3 Social

- The advanced training of Bertschi drivers during the implementation of the concept is one of the main social benefits that has been identified, as it contributed to the individual education of each driver.

10.2.7 Transferability

10.2.7.1 Transferables

Similar concepts have been implemented in collaboration with other large producers of chemical substances. Although this concept was developed especially for the chemical industry, the basic approaches could be transferred to other industries.

10.2.7.2 Limitations

Since the concept was developed as a tailor-made solution by Bertschi for Shell, difficulties may appear when attempting to transfer every last detail of the approach. One reason is that some producing companies still refuse to open their books to LSPs. Furthermore, the required geographic proximity sets spatial limits for the implementation of the concept.

10.3 Full Speed Supply Chain: Famosa – Full Speed Supply Chain

Sonia Guerola Pérez

By mid-2003, Famosa was a well-recognized brand, and seen by consumers as an innovative company, but its back-office logistics systems still dated from the 1980s. In the 1990s, Famosa had two main production locations, one in China and the other in Spain. The two were not coordinated, and just one distribution centre supplied goods to global markets. As a result, Famosa could not deliver goods on time to satisfy customer demand. The company decided to redesign its supply chain, starting in 2003. The final result was to achieve a "Speed Supply Chain" where Famosa is capable of tracking goods movements at any time and in any location, and of knowing what all the agents involved in the supply chain are doing. The new supply chain integrates physical flows together with information flows.

10.3.1 Problem Description

Famosa is a toy manufacturer and distribution company. Toys are strongly season products, a factor that impacts on the entire supply chain because sales are concentrated within 3 or 4 months. In addition, toy life cycles are very short; toys arrive at sales points at the beginning of October and have become obsolete by the end of December.

Due to this strong seasonality, the manufacturing process needs to be "stock to order", which implies high risks in stocking inventory. In addition, since China is its largest supplier and lead times are long, Famosa is dependent on the intensive use of forecasting.

Even though Famosa's supply chain was very short, it was very imbalanced. The company compiled a list of the resulting problems that required solutions:

- The company had a very high level of capacity whose utilization generated huge semi-finished inventories
- It had no structured production plan
- Operational and logistics costs were too high
- Products suffered high rates of obsolescence
- Famosa had established outsourcing in China, but used the country only as a source of components; products were still finished in Spain

- Famosa used to despatch products packed loose, without using pallets
- Production in Spain gave flexibility but incurred high cost inefficiency
- Distribution was carried out from a single point only.

The company could not deliver customer orders effectively due to stock outs, ignorance of the existence of orders, lack of warehouse management and control tools, etc.

10.3.2 The Solution

Famosa decided to reorganize the supply chain, moving to implement a "Speed Supply Chain". The solution was based in four key propositions:

10.3.2.1 Changes in Organization and Information Technologies

The first action carried out was to create an independent organization to manage a global and integrated supply chain, this way the company would become more agile and more capable in decision making.

The next actions taken were: the integration of the procurement process into operations management, moving from simple price negotiation to negotiation of business rules; the integration of the operations in China into the company's operations management, to improve quality and the reliability of production planning; the creation of a planning and replenishment department; and implementation of SAP in all the company's affiliates over 5 months.

10.3.2.2 Reactive Planning Processes

The main objective was to reduce the variability of forecasting accuracy due to market uncertainty. This involved: the creation of a planning department and the transformation of speculative capacity, which involved purchase orders to enlarge capacity which ignored customer demand, into reactive capacity, which is more reliable and provides an environment with a larger degree of comfort regarding forecasts. This last initiative was divided into the following tasks; adding to reactive capacity through the adaptation of the supply chain to the production cycles of the suppliers in China; lowering the lead time of the "AAA" products (products classified as very important or relevant) to permit products to be launched with significant changes implemented as late as possible; giving priority to launch products which have a lower impact if forecasting proves to be wrong; and determining which products bear higher deviation costs and rewarding customers who confirm their purchasing intentions in advance.

10.3.2.3 Collaborative Planning with China

Famosa decided to produce 80% of its end product in China, but this did not simply involve transfer to China of the production process. The company's relationships with suppliers are unusually close; Famosa carries out production planning, and negotiates a Master Production Schedule with its given supplier, which launches its own MRP. Thus Famosa has at its disposal the capacity necessary during peak seasons and is able to postpone loadings through a more efficient planning and negotiation process.

10.3.2.4 Reorganization of Distribution Network

Total redesign of the distribution network involved the following action:

- Implementation of central and regional distribution centers. Goods are dispatched direct from the distribution centre in China to the market, and to the central distribution centre in Spain. Some goods are also despatched from Onil to important customers, and small orders from Madrid. Famosa operates warehouses in all the principal European markets.
- The creation of a consolidation centre in China. The objective was to take advantage of logistics and fiscal benefits, including the potential of consolidating products from different suppliers; the supply chain consolidation centre is an export bonded warehouse.
- Adaptation of logistics processes to consumer goods sector standards: coding and labelling, EDI messages, barcodes and RFID in SAP, and the use of pallets.

10.3.3 Challenges

Challenge 1: Decide which products to outsource to China.

Challenge 2: The marked seasonality of demand for this type of product (sales being concentrated in only 3/4 months of the year). The company has to deal with very high levels of demand uncertainty, which requires sometimes excessive "push production".

Challenge 3: Toys are very short lifecycle products.

Challenge 4: Production capacity in China is limited, and Chinese suppliers cannot deal with production peaks.

Challenge 5: Long lead times involve intensive forecasting.

Challenge 6: It proved hard to integrate suppliers due to cultural and language issues.

Challenge 7: Famosa did not realize the full complexity of its supply chain network.

10.3.4 Lessons Learnt and Success Factors

- Lesson 1: Collaborating with suppliers allows companies to be fast, agile and flexible to satisfy customer demand, increasing competitiveness.
- Lesson 2: Even though best logistics practice recommends a pull manufacturing process, Famosa defends push production – it need not be a synonym for inefficiency. Working with push production can be beneficial if the product and sector so demands.
- Lesson 3: Stock levels change with a unique warehouse storage centre or if goods are distributed from three centres due to the huge concentration of toy sales in 3 months of the year.
- Lesson 4: Even when products have a very marked seasonality, good planning practice can make working with a pool of pallets profitable.
- Success factor 1: Famosa strengthened its relationships with suppliers in China and improved supply chain reliability. The company's main objectives were coordinated with suppliers.
- Success factor 2: The company added two new processes to improve its working methodology: the introduction of forecasting, and improvements to packaging to make transportation easier.
- Success factor 3: Famosa introduced the use of SAP across its whole supply chain.
- Success factor 4: Famosa employees had to learn how to use SAP and were trained in operating the software. They also had to gain familiarity with and know how work in the new supply chain and with the new processes.
- Success factor 5: In the beginning relationships with suppliers in China were difficult due to cultural differences and the classic relationships big companies have with their suppliers in Asia. Famosa tried to be as flexible as possible and adapt to suppliers' customs and improve the relationships, including their suppliers in the network and in the supply chain, and treating them as part of the company.

10.3.5 Benefits

10.3.5.1 Economic

Service levels have been improved, reducing late deliveries by 15%. The company's service level has improved by around 13% and nowadays Famosa is able to supply its customers with better production quality and higher volumes of stock available for delivery. Productivity has increased due to production of finished products in China and the opening of distribution and consolidation centres. Distribution costs have been reduced,by 4%, as have storage costs both for work in process and finished products (by 5.5% and 7%, respectively).

In the past Famosa's customers did not enjoy good access to the company's products; now its products are available more widely in shops, because Famosa can fulfill customer demand with a new, flexible supply chain capable of sending Famosa's products to any point in the globe.

10.3.5.2 Environmental

Famosa reduced obsolescence inventories, thereby reducing product waste.

10.3.5.3 Social

The number of employees in Spain and in the other countries in which Famosa operates has increased. Some of its workers have learned how to use SAP and to manage international commerce. Some of the managers have also learned about their supply chain network.

10.3.6 Transferability

10.3.6.1 Transferables

Many businesses have already developed this type of strategy, but they tend to be multinationals and very large companies. More and more companies, like Famosa, will need to make changes in their supply chains to improve profitability, and even to remain in business, no matter which country they are based in, or which sector they belong to.

10.3.6.2 Limitations

This practice can be transferred to any kind of company; the only restriction is its size of it. Big companies can afford to invest money in a strategic change such as Famosa's programme, but for small companies it is very difficult to create such a big supply chain. The design of the Famosa project has no geographical or industry-sector limitations.

10.4 Collaborative Intermodal Transport with Supplier: Ikea and Com40 – Reconfiguration of the Supply Chain Structure

Barbara Ocicka under supervision of Krzysztof Rutkowski

Due to copyright issues this case is exclusively available online at http://www.bestlog.org. The case can be found in the "knowledge Base" section at the website.

10.5 Pooling: FM Logistics – Pooling in the Retail FMCG Sector

Katarzyna Gapska under supervision of Krzysztof Rutkowski

Due to copyright issues this case is exclusively available online at http://www.bestlog.org. The case can be found in the "knowledge Base" section at the website.

Chapter 11
Processes and Operations Cases

11.1 Increased Transport Efficiency: IKEA – Increased Transport Efficiency by Product and Packaging Redesign

Gunnar Stefansson

In order to lower logistics costs and increase efficiency in its transportation and warehousing operations, IKEA started an internal competition to reduce unnecessary air in their product packaging. This "Air hunting competition" focused on removing as much air as possible from packaging and thereby increasing true product volume during transportation and storage. Several IKEA products were identified for packaging development, one being the Glimma tea candle that is described in this best practice documentation. The development of the Glimma tea candle packaging resulted in a 30% increase in products volume for each load unit. Thanks to this packaging development, the efficiency of the transportation and warehouse operations is now much greater and the impact on the environment has decreased significantly.

11.1.1 Problem Description

Many of IKEA's products are low value but high volume products. Transportation costs form a large part of the total cost of many of the products, which makes it important for the company to minimize transportation, handling and warehouse costs wherever possible.

A problem IKEA has identified is that many of their low value products use excessive space on load units due to packaging design. This packaging design results in poor efficiency in both the transportation and warehouse operations, and thereby creates unnecessary costs. Examples of large products are pillows, mattresses, sofas, and others which are filled with air and therefore take up a lot of

C. Tyssen et al., *Sustainable Supply Chain Management*,
DOI 10.1007/978-3-642-12023-7_11, © Springer-Verlag Berlin Heidelberg 2011

unnecessary space during transportation and storage. Another example of a product that has traditionally included a lot of air in its packaging is the Glimma tea candle. The original packaging for the candles is a bag containing 100 candles weighing in total 1.4 kg.

IKEA has been looking into ways of eliminating as much air as possible from their product packages. By doing this the utilization of transport load units and vehicles would be increased, and thus lead to higher efficiency and lower transportation, handling and warehouse operations costs.

The excessive packaging space used during transportation affects the environment as well, as too many vehicles are being used for transportation operation.

Solving the problems affects packaging design and even in some cases product design, as is has to be adapted to the new packaging design. IKEA's suppliers need to be involved in decisions on packaging design, as the manufacturing processes are often affected.

11.1.2 The Solution

The Glimma tea candles' original packaging was a plastic bag randomly filled with 100 candles. The round shape of the candles and the nature of the packaging process resulted in a bag containing a substantial portion of air. To solve this problem, the air has to been taken out of the bag and the number of candles per unit increased. The solution is to place the candles in such a pattern that the density of the product package is increased, removing excess air. Both of these solutions depend largely on the product's shape and the potential to improve the package design.

By using a new packaging method it is possible to decrease the amount of air enclosed in the Glimma packaging process. Instead of using the previous method of randomly filling a plastic bag with 100 Glimma tea candles, they are now neatly stacked in five rows, 4 × 5 candles in each layer. The result was that more candles could be loaded on each pallet, rising from around 250–360 packs, resulting in a near 30% increase of products per load unit.

Increasing the density of the packaging led to an increase in weight per pallet. Previously about 60,000 pallets in total were shipped worldwide every year. By increasing the load capacity by 30%, the total amount of pallets could be reduced to 42,000 pallets worldwide. This results in a significant decrease in transportation work and costs.

However, the weight of the load unit increases to such a level that the weight can exceed the load capacity of vehicles if they loaded to reach full volume utilization. The weight of a 40 ft container with full load would be approximately 22,000 kg, which exceeds the maximum weight allowed for a vehicle, semi-trailer and load in some European countries.

The solution to this overload is to balance the load on the trucks by using lightweight products to fill up the left over space. This was accomplished with voluminous but lightweight products such as mattresses and folded sofas. This load balancing results in high utilization, in respect of to both weight and volume.

To be able to achieve this type of balancing, products need to be consolidated from different suppliers and travel together to a common destination. IKEA has solved this by using a so-called "cluster supplier". The "cluster supplier" concept is based on a solution where a major supplier of IKEA takes responsibility for storing goods from the other suppliers for later consolidation with its own goods when orders are despatched to central distribution centres and warehouses in different countries. For the Glimma product, the candles are sent to a cluster supplier which in some cases is also a supplier of mattresses.

The cluster supplier concept typically involves several smaller suppliers, from seven up to 24, in the area around the cluster supplier, who deliver their products direct to the cluster supplier. When orders are placed all the shipments are consolidated and sent to the ultimate destination. There are four suppliers of the Glimma tea candle worldwide, located in Europe as well as in Asia.

11.1.3 Strategic Implementation and Continuity

It was very important for IKEA to work closely with the suppliers, as in many instances the packaging processes needed to be changed. In some instances considerable investments needed to be made and as the total cost of the products was not to be increased, the costs needed to be absorbed through increased process efficiency.

Another important aspect of the implementation was to fit the Glimma distribution to the IKEA "cluster supplier" concept. The candles need to be transported to a cluster supplier that can then consolidate the relatively heavy candles with lighter products such as pillows, mattresses and the like. It was critical to the success of the project to achieve this, or the benefit of reducing the air content in the product packaging could not be fully realized. This is due to being unable to fill the transportation units fully due to weight constraints. By integrating flows from several suppliers, it is possible to consolidate goods of different density to balance the transportation weight and volume ratio.

11.1.4 Challenges

Challenge 1: Finding ways to reduce unnecessary space in packaging without having to alter the product design too much.

Challenge 2: Load balancing products of different volumes and weights to fill transportation units, in regard to both weight and volume.

Challenge 3: Motivating suppliers to take part in the development as in some instances the production processes have proved more expensive.

Challenge 4: Stick to the original price of products, although investments needed to be made in new packaging technology.

Challenge 5: Develop a packaging method to ensure as little air remains in the package as possible.

Challenge 6: Integration of various suppliers in an area within a "cluster supplier" scenario, as the innovation involves storage of one company's products by another supplier.

11.1.5 Lessons Learnt and Success Factors

- The solution is itself not innovative, but it shows how greatly small changes can affect the utilization of transportation and warehouse resources. At times, small but continuous improvements will yield better overall improvements and cost benefits than a small number of large projects that are more likely to fail to achieve their goals. Over time this type of thinking could be applied to other products as well, to lower costs and increase both economical and environmental efficiency, mainly in transportation operations.
- It is important to find new ways of integrating the various suppliers in the supply chain to be able to improve efficiency and lower costs. Not every company would be able to integrate suppliers in this manner, but companies of IKEA's size are more likely to be able to negotiate agreements with their suppliers such as the cluster solution.
- Small and relatively simple changes to something such as product packaging can lead to great improvements in efficiency and cost efficiency. However, even though it is a relatively small change in a product's packaging, it can still lead to investments in new packaging equipment to be able to fulfill the project requirements. In that case it is important to consider whether the benefits of the investments are indeed greater than the costs.
- It is of the utmost importance to involve suppliers in the development at an early stage, preferably with people who have hands-on experience of the product and can contribute good ideas in line with the project goal. In most cases contributions from the business partners involved are more likely to be successful than directives coming from the customer.
- Unforeseen side-effects are not uncommon and in the Glimma case, the packaging technology led to much more convenient "shop ready" packaging. The design means much less time is spent in preparing the product for display in the retail warehouses.

11.1.6 Benefits

The major benefit is much more product in each load unit, meaning better utilization of transport units and storage space. In addition, handling becomes much easier as fewer units are being handled, both at supplier sites, central distribution centres and in warehouses. Thanks to increased efficiency in transportation and handling, the impact on the supply chain and the environment is decreased.

11.1.6.1 Economic

- Utilization of transportation resources increased by 30% for both IKEA and its business partners
- More space in warehouses that can be used for other purposes, such as storage or display of other products
- The new packaging makes it easier for the customer to handle the product in the store
- Easier to replenish the warehouse as the product is more or less "shop ready"

11.1.6.2 Environmental

- Less noise and emissions due to approximately 400 fewer 40-ft containers on the roads.
- Reduced need for warehouse space.
- Smaller packaging leads to use of less packaging materials (corrugated paper and plastic film).

11.1.6.3 Social

- 400 fewer trucks less on the roads reduces the risk of accidents (no precise figure is available)
- Less transportation, however, affects employment within the transportation industry

11.1.7 Transferability

11.1.7.1 Transferables

Eliminating empty space (air) from packaging means that the unit volume is decreased, and this can be achieved for most products in most industries, independent of geographic location and company size.

11.1.7.2 Limitations

Investments are needed in production modifications.

11.2 Transportation of Healthcare Products by Inland Navigation: Baxter – Transportation of Healthcare Products by Inland Navigation

Gunnar Stefansson

Baxter uses inland navigation to transport many of its medical and biotechnical products from the deep sea areas of the port of Rotterdam and Antwerp to their distribution center located in Lessines, Belgium, rather than traditional road transportation. The inland navigation solution is also used for the transport of export of export products from the European distribution center to regional distribution centers around the world using the same containers, thereby increasing the utilization degree of the containers and decreasing empty runs.

11.2.1 Problem Description

Reliability is the main concern for Baxter in its worldwide healthcare business. The reasons are the strict deadlines and high-level requirements of the healthcare sector. Uncertainties in transportation and delivery processes are not accepted, either by customers or by Baxter itself.

One of the long-time problems has been the uncertainty in delivery times from the large ports in Europe such as Rotterdam and Antwerp due to variations in the availability of services at the port and congestion on the roads outside the port area. The port in Antwerp is 111 km from Baxter's European distribution center at Lessines, and Rotterdam is 215 km away. There is a high risk of congestion on the relevant roads, especially close to the ports and around Brussels.

This situation makes it very difficult for the company to ensure reliable deliveries from the deep sea part of the ports to the distribution center in the hinterland. The variations in delivery time can be anywhere between 8 h and 3 whole days. At the same time, congestion on the roads and around the ports makes it difficult to deliver export goods to the ports in time for the departing ships in the company's distribution system for products leaving the European distribution center in Belgium and heading for distribution centers in other areas of Europe and the world.

11.2.2 The Solution

Baxter's supply chain management system coordinates products from tens of thousands of suppliers worldwide to the various Baxter manufacturing facilities around the world, developing manufacturing schedules to meet forecasted demand

and inventory needs and distributing finished products to customers in more than 100 countries. Distribution reliability is very important for the company, and it relies on its supply chain to deliver the right products at the right time.

In the search for more reliable transportation solutions in Europe, Baxter looked at inland navigation alternatives in the Netherlands and Belgium region, where high volumes of goods need to be transported to and from its European distribution center (EDC) in Lessines near Brussels. The inland navigation alternative turned out to be feasible once business relations with established with suitable service providers in the area.

Baxter uses inland navigation to transport many of its medical and biotechnical products within Europe rather than relying on the traditional method of road transportation. Baxter introduced the use of inland waterways from the deep sea areas of the ports in Antwerp and Rotterdam in the mid nineties and today (2007) organizes the shipment of more than 1,000 containers from the ports to the European distribution center each year.

The distance from the port of Rotterdam to the European distribution center in Lessines is 215 km, while Antwerp is 111 km away. The inland navigation solution includes barge transportation from the deep sea port areas where the barges are loaded to the Avelgem Container Terminal located 34 km from the EDC. The delivery times from Antwerp and Rotterdam are 18 and 14 h, respectively, which allows 24-h service (on average). The containers are transported from the Avelgem Container Terminal by truck.

After the success of the solution with imports, Baxter is now also using the inland waterways distribution platform for the transport of its exports as well as transporting products from the distribution center to regional distribution centers in Europe and around the world. The containers used for imports are turned around and used for export, increasing the utilization degree of the containers and reducing the number of empty runs.

The solution has turned out to be a success in various ways. Costare down 40% compared to previous solutions. At the same time, delivery reliability has increased as delivery variations have decreased.

The environmental benefits are obvious, as barges only consume 20% of the fuel needed to transport each kilo of goods by truck. An average ship can carry as much as 120 trucks with a 40-foot container loading space, and this takes a high volume of truck traffic off the roads.

11.2.3 Strategic Implementation and Continuity

As long-term relationships have been established with suitable partners in the inland waterway networks, all the parties involved are committed to developing the solution and ensuring that it will work in the long term.

Due to the reduced environmental impact, the solution is sustainable. Moreover, the solution is in line with Baxter's strategy of the "green supply chain" geared towards the continuous reduction of the carbon footprint of its products.

In addition, the solution is not only less costly than earlier transportation solutions but also and more reliable, paving the way for increased use of inland waterways in the future.

11.2.4 Challenges

Challenge 1: To persuade transportation and logistics companies to participate in the setup, both at the deep sea port area and the inland container terminal. It was necessary to establish services in the deep sea ports complete with the option of loading on river boats or barges.

Challenge 2: To ensure that transportation costs would not exceed the previous cost of road transportation.

Challenge 3: To ensure that total transit time would not increase with the new solution.

Challenge 4: To find a service provider to set up a stock yard for this solution.

11.2.5 Lessons Learnt and Success Factors

- Inland navigation has traditionally been used for the transportation of low-value goods such as bulk products or project cargo. The transportation of high-quality goods via inland waterways is, however, just as effective.
- The reliability of inland navigation is superior to that of road transportation and little or nothing is sacrificed in terms of flexibility despite the obvious limitation due to the dependence on infrastructure.
- Increased reliability was the main driver behind the solution, and the desire to minimize variations in delivery time. At the same time, the new inland navigation solution reduced total transportation costs by 40%, a welcome outcome but not one that was prioritized at the outset.

11.2.6 Benefits

11.2.6.1 Economic

The solution has improved service quality as delays have decreased. Customer service levels have increased due to improved delivery reliability. Product availability in the European distribution center has increased thanks to improved inventory monitoring and control. Easier planning of distribution activities is a further benefit. The containers used for imports are also used for exports, thereby reducing the number of empty runs and increasing resource utilization. The efficiency of warehouse personnel deployment has increased, as the greater reliability of deliveries from the ports and more precise information on the progress of transportation makes planning easier. Transportation costs have been reduced by up to 40%. The planning of activities is easier and the visibility of activities is better. Improved product availability ensures effective response to customer needs. Products can be delivered to the customer faster in the event of a sudden increase in demand.

11.2.6.2 Environmental

The inland navigation solution has lowered CO_2 emissions, as there are around 500 fewer trucks on the road each year. The river boats use only 20% of the fuel to move one unit a certain distance compared to road transportation. This impacts land use, as there is a shift in the mode of transport and a resulting need for new inland container terminal space. At the same time, however, space requirements in the ports can be slightly reduced, as some containers can be stored at the inland terminal rather than in the port area.

11.2.6.3 Social

With more than 500 fewer trucks on the road each year, the likelihood of accidents is reduced – and inland navigation accidents are not likely to increase because of the solution used by Baxter. Employment levels have been affected: there is not as much demand for truck drivers but the need for inland waterway personnel has increased at the same time. The effects are not directly comparable, however: one river boat can carry between 100 and 500 containers (TEU), making far more effective use of personnel resources per load unit than trucks (which can only carry two units).

11.2.7 Transferability

11.2.7.1 Transferables

Using inland navigation for the movement of high-quality containerized goods is a suitable concept for all kinds of goods from various industries that use containers in their transportation solutions. The solution is applicable in all geographic regions that have inland waterways and an infrastructure of inland container terminals with effective links to the road infrastructure covering the final stretch to the end customer.

11.2.7.2 Limitations

Inland navigation has great transferability but there are some limitations. The main constraint is access to infrastructure including ports, inland waterways and inland container terminals – as well as road, if the company has no direct access to the terminal. In addition, service providers providing the necessary services must be in place, and the volume of goods must be high enough to make the solution economically viable.

11.3 Intermodal Transport in Retail: Migros – Boosting Rail Transport

Christoph Tyssen

In 2005, Migros, a diversified holding from Switzerland was facing a serious price war, especially in its retail business. This was mainly due to the entry of discount oriented competitors into the Swiss retail market. Those two competitors followed a significantly aggressive pricing strategy, putting the actors established in the market under intense pressure. In order to cope with this intensifying competition successfully, Migros set an objective of lowering transportation cost by 10%. To reach this objective, the company developed a nine point agenda. As a consequence of this nine point agenda, the rate of rail transportation was increased significantly.

11.3.1 Problem Description

Migros introduced a new logistics strategy for the entire company. As a part of this strategy, transportation costs were examined. Additionally, in 2005, Migros was facing an increase in transportation tolls and transportation costs in terms of rising fuel prices and augmented road tolls for trucks. Additionally low-price retailers entered the Swiss market, intensifying the price war in the market. As part of the logistics strategy, the company developed a concept to reduce transportation costs and keep its market position. Hence, the concept NATRO was developed to lower the transportation costs by 10%. After analyzing the organisations cost structure, the reduction of transportation cost was one identified lever to achieve this goal. Hence, the agenda called NATRO consisting of nine points, was developed. The action plan aimed at an optimisation of Migro's regional and national transports as well as taking over the procurement hauling, forming the three pillars of the new concept. Since the description and analysis of all nine concepts introduced into the company would go beyond the scope of this case, one of the most significant action, the intensified use of the railway system, and its consequences will be described in the following. Both were primarily owed to on the one hand the reduction of the frequency of delivery and on the other hand the optimization of direct-delivery.

11.3.2 The Solution

Migros was facing risen costs in transportation and an intensified competition in the swiss retail market due to low price competitors. This development created a sense of urgency, to act and to stay competitive, within Migros. In order to be successful regarding this point, the company defined the clear goal lowering the level of transportation cost by ten percent. Following, several scenarios were developed, aiming at providing a wide range of options for the company's management. Some of these scenarios were more aggressive, while others envisioned a rather defensive corporate strategy. After a careful analysis and evaluation of each scenario, Migros decided to choose a nine-point agenda to tackle the urgent challenges. One of the actions defined by this "schedule" was the optimisation of the organisation's national distribution processes. This action was to be realised on a regional, as well as on a national level. Closely connected to this point was the intensified use of the train as a means of transportation. The reasons for this decision were multisided. Simultaneously to defining the goal of a ten percent transportation cost reduction, Migros' management clearly communicated, that the quality for the customer should not suffer from the actions taken connected to this objective. When restructuring transportation processes and delivery concepts, the various KPIs in retailing should not suffer. One of the KPIs in retailing, measuring quality for the customer is the on-shelf-availability, i.e. the right amount of stock in the right place at the right time. In order to ensure excellence regarding this point, Migros had to guarantee in-time deliveries to each of its stores. Though out of stocks are not necessarily connected to the transportation mode one needs to bear in mind to keep those objectives considered when implementing a new transportation programme.

Nevertheless, the intensified use of Switzerland's sophisticated and highly reliable railway system needed some preparation. As part of the NATRO concept, transportation volumes for railway transportation were raised. This also inured to the benefit of railroad transportation, as this transportation mode is highly fix cost connected. In order to realise cost advantages by using this concept, Migros had to adapt its transportation volumes to the relatively high fixed costs connected to train transportation. As rail transportation is highly fixed cost dependant the only way of lowering transportation costs and receive better price rates is the augmentation of transported volume. Migros was in a good position to do so, as the company operates with two distribution centres for entire Switzerland. As rail transportation is especially worthwhile for long distance hauling, Migros thereby could shift more goods on long haul rail transportation. Another benefit was the existing infrastructure for rail transportation. As part of the company's tradition, Migros' distribution centres were already equipped with a railway siding. Additionally all suppliers possessed a railway siding as well. This altogether lead to higher volumes of goods on rail without educing negative effects like long some transportation or many transshipments processes.

Besides the increased volume shifted to rail, Migros worked closely together with its suppliers. The company managed to negotiate a procedure, which made stock-keeping and time management for the organisation more efficient. Instead of storing the procured goods at the producer and eventually in Migros' facilities, the company managed to imply a system, which would allow them to pick up finished goods at the supplier's factory and ship it directly to its distribution centers. This achievement made Migros' procurement process much leaner and quicker.

Nevertheless, due to the high complexity in the organisation's procurement and distribution processes and the urge to align national and regional interests, a cleaner and more efficient transportation system was established. This led to less costs and environmental transport as well as providing the same service level as before.

11.3.3 Challenges

Challenge 1: Imply change management: One of the main challenges, for reaching the 10% reduction in transportation cost, was to realise a certain degree of change in the system of Migros. The company's regional departments play a strong role, while enforcing their interests face to face with the holding organisation. This system was developed over decades. Hence, the structures are quiet developed and established. Introducing innovations into this system of regional and national interests was quite a challenge.

Challenge 2: Consider the effects of change: As mentioned above, the system of procurement and distribution of Migros was well established. Hence, the changes planned and their effects on the overall system of Migros needed to be analysed were carefully in order to avoid significant negative effects.

11.3.4 Lessons Learnt

- Realise the importance of managing politics within your company, i.e. the alignment of national and regional interests
- Map different scenarios to stay flexible
  - Compare the scenarios
  - Be aware of the respective consequences of their application
    Adapt your behaviour to the structure of your organization
    Carefully analyse the effects of changes in a well-rehearsed system and aim at avoiding negative effects
    Treat your suppliers well and aim at collaborative solutions

11.3.5 Success Factors

- Define clear goals (here: cost reduction) before mapping different approaches
- Clearly communicate your goals within the company, i.e. guarantee that everybody in the organisation is "on the same page"
- Maintain the success factors of your company, while realising optimisation projects
  - Keep quality for the customer on a constant level or even improve it
  - Ensure the punctual delivery of goods, while decreasing the number of deliveries
- Take advantage of your bargaining power
  - Map the requirements for the restructuring of your organisation and the respective bargaining opportunities you have
  - Here: Migros could shift more goods to the railway and was thereby able to negotiate better price rates for the rail transport.
- Cooperation in order to consolidate transport runs

11.3.6 Benefits

The intensified use of the railway system lead to several benefits. The main benefit from an economic point of view was the reduction in the share of transportation cost. More goods were moved by train, while transportation cost increased only slightly. A further benefit from the increased transportation of goods by train is the obvious reduction in CO_2. Since Migros owns 400 vehicles used for the transportation of goods, the intense use of the railway system contributes indirectly to an improvement of the traffic holdup issues on Switzerland's roads. Overall, it can be stated that occurring benefits outweigh the negative impacts of Migro's action plan.

11.3.6.1 Economic

- A decrease in the share of transportation cost was realised, since a larger amount of goods is transported by train.
- The approach of moving goods mainly by train also increased the volume, which is transported per sending.
- Due to traffic holdup issues on Switzerland's roads, the transportation by train offered Migros a higher reliability.
- The shelf availability could be maintained.

11.3.6.2 Environmental

- Obviously, the transportation by train reduced the amount of CO_2 and other toxic emissions caused by goods transportation.

11.3.6.3 Social

- Despite Migros shifting a significant amount of its transportation to the railway, the company managed to maintain the majority of the drivers employed by the company.
- Less congestion on the road and less annoyance of car drivers by less truck on the road.

11.3.7 Transferability

11.3.7.1 Transferables

Generally, the shift from the road to the train can be seen as an approach, which is transferable to other industries (e.g. Chemicals). If planned well, this solution will create cost efficiency effects for the companies involved in the process. Moreover, this approach is able to realize a positive impact on the perception in public.

11.3.7.2 Limitations

Despite its benefits, this approach has some limitations to it. The amount transported on the tracks needs to be sufficient in order to create a cost advantage compared to road transportation. Other possible limitations to the increased use of the train are the bargaining power of the company, as well the nature of the respective country's geography.

11.4 Intermodal Less Than Truckload: Cargo Domizil – Intermodal Less Than Truckload Transport

Christoph Tyssen and Thorsten Klaas-Wissing

The Swiss logistics service provider Cargo Domizil offers unit load shipments by combined road and rail transport. Long haul transports, usually executed via road transport, are performed by rail, offering a better level of service and cleaner transport than road-only services. The major challenge for Cargo Domizil's management was to turn this formerly state-owned loss-making business into a profitable private company, and this required major restructuring. Since the shareholders of Cargo Domizil are competitors in other business sectors, coordination between the enterprises associated with the company was a major issue for the management –

especially since they were working together without any complex contractual arrangements.

11.4.1 Problem Description

Cargo Domizil was launched in 1981 as a division of SBB (the Swiss railways) to execute unit load rail shipments. As the aim of the Swiss government and of the state-owned SBB was to shift freight traffic from road to rail, SBB developed long-haul rail freight services with collection/distribution by truck. The business model initially failed due to organizational and operational inefficiencies, and the operation was unprofitable.

Meanwhile, private sector transportation companies in Switzerland were affected by a change in the government's transport policy – specifically by a national ban on night-time trucking (10 pm–5 am) and by relatively high per-kilometer tolls for trucks on all Swiss roads.

After its partial privatization in 1992, SBB sold Cargo Domizil to a consortium of three large and medium-sized road haulage contractors. This consortium is called Transvision and comprises Camion Transport AG, Galliker Transport AG and, Planzer Transport AG. These co-owners needed to restructure Cargo Domizil's operations to make its "combined mode" service competitive with road haulage services, capitalizing on the advantages of overnight rail movements and the negative image of road transport, especially in Switzerland.

Cargo Domizil uses trains for long haul transport within Switzerland partly to avoid the night-time trucking ban.

11.4.2 The Solution

Cargo Domizil offers less-than-full-truckload (LTL) shipments via combined rail/road facilities. This system is seen as unique in Europe, where LTL shipments are generally moved solely by road. Daily 24-h deliveries of unit loads to locations across Switzerland are guaranteed by the road/rail network. The three partner companies manage 11 Cargo Domizil logistics depots between them, and all depots are at rail trackside locations.

Cargo Domizil trucks collect consignments during the afternoon and take them to these rail-side depots, where they are loaded on trains, forwarded to Olten near Zurich, sorted by the SBB and sent to their destination depot overnight. (Cargo Domizil normally uses around 250 rail cars per night). In the morning, trains arrive at the destination logistics depots, where the freight is transferred back to a truck and delivered to the customer in the morning.

Besides avoiding the night-time trucking ban and expensive road tolls, the use of rail offers significant advantages in Switzerland's mountainous regions, which are sometimes difficult to reach by truck (Switzerland built tunnels exclusively for trains).

Thanks to the hub-and-spoke structure of Cargo Domizil's logistics depots, the company's drivers always operate in the same region and are familiar both with routes and individual customers – an advantage in providing courteous and friendly service, especially in Switzerland, which has three official languages and where customers expect to be addressed in their own mother tongue.

In order to cut fixed costs and make Cargo Domizil profitable, the number of logistics depots had to be significantly reduced without affecting 24-h delivery services throughout the network. This was achieved after the Transvision consortium reorganised some of its collection/delivery routes and negotiated an agreement with the SBB Cargo division of the Swiss railways that ensures adherence to delivery times and competitive rail transport rates.

Structural changes were accompanied by a training program, as Cargo Domizil considered that its employees were the key factor in efficient processes and on-time-deliveries.

Since the three consortium partners were also competitors and had to manage their own logistics depots, effective communication was of major importance and common IT systems needed to be implemented.

The consortium partners also needed to find practical ways to avoid inequalities in the allocation of Cargo Domizil's profits. Transport volumes vary within Switzerland; some regions send more goods outwards than others, while others receive more shipments than they send, resulting in different financial flows in different areas. Adequate definitions of value had to be agreed upon in order to ensure equal allocation of revenues.

11.4.3 Challenges

Challenge 1: Turning around an unprofitable state-owned entity to become a profitable less-than-truckload service provider

Challenge 2: Implementation of a business model exploiting the advantages of rail

Challenge 3: Reducing the handling time for unit load transfers between trucks and rail wagons

Challenge 4: Designing an optimum, future-proof structure for a rationalized logistics center network proved complex

Challenge 5: Implementing collaborative behavior within a consortium of competitors

Challenge 6: Cooperating without a complex contractual arrangement, keeping formal rules to a minimum

11.4.4 Lessons Learnt and Success Factors

The success of the restructuring of a formerly state-owned company was underpinned by the collaboration-based relationship between the consortium partners. Only their strong commitment could have ensured Cargo Domizil's continuing success. The partners are still working together without any detailed contractual arrangement, other than the articles of association of Cargo Domizil AG itself.

- The consortium realized early on that logistics customers preferred "green" transport to conventional road haulage, but only when offered at the same price, so it wanted to offer rail transport services for unit loads.
- The consortium wanted to gain a competitive advantage over other logistics providers by moving goods at night when competitors could not use their trucks.
- Switzerland's extensive rail network and the partners' local distribution operations enabled them to cover the whole of the country while reducing the number of own distribution depots.
- New operating structures were particularly important in enabling the success of the now privately-owned Cargo Domizil.
- Personnel training was particularly important, given the management and coordination challenges created by reducing the number of logistics depots.
- Last but not least, the political environment in Switzerland had a major influence on the success of the concept of LTL rail shipments. The night-time trucking ban and high road tolls in combination with Switzerland's highly developed rail network contributed significantly to the project's success.

11.4.5 Benefits

11.4.5.1 Economic

- As each shipment is delivered by the same person, customers across Switzerland enjoy friendly service from an individual sharing their own cultural background.
- Increased capacity utilization through consolidation of loads at depots: competitive advantage as the concept bypasses the Swiss ban on night-time trucking
- Competitive advantage of green labeling

11.4.5.2 Environmental

- Rail is considered less CO_2 emission-intensive than road freight transport
- Fewer fine particulates are emitted by electrified rail transport, as energy can be produced with less pollution in power generation plants than in truck engines

- Electricity used by trains can be generated from alternative energy sources
- Less energy per ton-kilometer is used by trains than by trucks

11.4.5.3 Social

- Lower transport emissions due to use of rail
- Rail freight is generally safer than road haulage; a single train handles more cargo than a single truck, thereby reducing the risk of accidents in proportion to the tonnage carried
- Specialist training programs are available for intermodal unit cargo handling
- A sustainable business model secures jobs in the Swiss rail freight industry
- A source of noise was removed from the roads (but shifted to the railway)

11.4.6 Transferability

11.4.6.1 Transferables

The Cargo Domizil solution is transferable to a company of any size. In particular, larger companies with their own logistics infrastructure networks could easily adopt the concept of transporting unit loads by train overnight. The concept could easily be adopted, and possibly enhanced, in other countries offering similar logistics environments beyond Switzerland.

11.4.6.2 Limitations

The particular political environment in Switzerland played a major role in the success of the Cargo Domizil concept. It would therefore not be easily transferable to other countries with significantly different transport policies or infrastructures. A further limitation is that the concept is feasible only for logistics service providers supplying less than full truck unit load services.

11.5 Collaborative Intermodal Transport with LSP: Mercadona and Renfe – Intermodal Collaboration Distribution

Sonia Guerola-Pérez

Mercadona, with Acotral and Renfe, developed a plan to promote sustainable transportation in Spain, and signed a contract with Acotral and Renfe for them to transport non-fresh food and non-food goods. The contract specified that Renfe had to provide eight trains a week from Sevilla to Tarragona and from Sevilla and Valencia in a round trip. This new route connects suppliers' warehouses in the south of Spain with Mercadona Distribution Centres in the east coast. This solution enabled Mercadona to reduce CO_2 emissions by over 12,000 tonnes due to the number of its truck deliveries being reduced by up to 9,152.

11.5.1 Problem Description

Mercadona is a Spanish distribution company in the supermarket sector. This company transports 19,000 tonnes of goods daily from the eastern coast to southern Spain.

The company is sustainability-oriented and wanted to their suppliers to join in pursuing the company's objectives to obtain a win–win result. Taking into account these primary objectives and environmental concerns being inherent in Mercadona's policies, the company developed and launched an enviromental plan involving its main logistics and transport supplier, called Acotral.

Mercadona transport strategy is intermodal oriented. It is inside of the sustainable company policy. Road transport is congested in Spain and in Europe, being the most commonly used mode of goods transport.

On the other hand, Renfe is a train company carrying both passengers and goods. It had suffered losses following market liberalization and needed to change their market positioning in order to fight back in this unexpected situation. As a strategic target, Renfe wanted big supermarket chains to see rail as an effective transport alternative.

Mercadona being a sustanaibility-oriented company, it also wished to shift mode from road to rail.

11.5.2 The Solution

Mercadona, Acotral and Renfe-Freight signed a contract continuing until 2010 to transport non-fresh food and non-food goods between Sevilla and Tarragona/Valencia. Under this contract, they have a collaboration agreement whereby Mercadona relies on Renfe to deliver the goods on time and Acotral is responsible for the logistics management and the transport coordination. With this contract Mercadona increases its sustainability.

Both supplier companies coordinated their efforts. Acotral scheduled the trucks to pick up the goods from the trucks at the specified times, and Renfe fulfilled the promise of delivering them on time.

The principal feature of the innovation for Mercadona is its collaborative agreement wherein it relies on Renfe. Secondly, it is uncommon for a customer to involve one of their transportation and logistics providers in improve their efficiency performance and sustainability profile.

Goods are transported by truck from suppliers in Sevilla to the rail terminal where the trucks are unloaded and the train is loaded with the goods.

This process is not easy because queues to load and unload goods are very common at suppliers. As Mercadona is very concerned with the reliability of this transport chain, Renfe gives preference to these Mercadona's loadings.

Mercadona had no previous experience of intermodal road/rail transport, so it was important to learn about the system moving to the new alternative.

Once it is loaded, the train transports the goods to a rail terminal in Tarragona/Valencia where the truck company (Acotral) picks up the goods again and takes them to the Mercadona distribution center.

Renfe has developed a communications platform to track goods transported in its trains. Mercadona has also integrated its own communication system with Renfe's to ensure visibility through the whole supply chain. Train journeys take place twice a week, in two trains Renfe that has assigned exclusively to Mercadona. The total load transported during a complete year is 220,000 tonnes. Mercadona has found improvements in goods being delivered reliably on time, and at lower cost, in a more sustainable way.

This solution permitted Mercadona to cut up to 9,152 truck delivery journeys and to reduce its freight transport CO_2 emissions by over 12,000 tonnes.

11.5.3 Challenges

Challenge 1: Coordination between the road transport and train companies to manage the logistics chain.

Challenge 2: Coordination between trucks and the warehouses to achieve on-time fulfilment of supermarket deliveries.

Challenge 3: Customisation of Renfe freight services to fulfil Mercadona's requirements, with two trains per week assigned exclusively to Mercadona.

11.5.4 Lessons Learnt and Success Factors

- Lesson 1: Nowadays rail transport is not an economic alternative for short and medium distances. It is good for long distances. On the other hand, it is a new and effective solution for consumer goods distribution.
- Lesson 2: It is hard to implement a collaborative system. The collaboration among partners should be based on trust and on a win-win scenario.
- Lesson 3: It was worth improving the company's logistics in terms of economic and environmental advantages. This is consequence of developing a new strategy to manage operations, prioritize loads, decide the types of goods transported, obtain energy savings, etc.

11.5.5 Strategic Implementation and Continuity

Railway transport for goods is not a unique or innovative practice, but what is innovative is the way that a customer or a large company enrols its transport and logistics provider in the search for a win–win solution. In addition Mercadona and its partners improved the sustainability of the client's transport.

This practice is easy for competitors to copy, the only obstacle is the availability of rail capacity. The key drivers and motivations for the different companies involved in this case include their commitment, the benefits, and the business growth achieved.

Mercadona has a strategic environmental commitment. This good logistics practice allows the company to reduce costs and be environmental and social responsible. In addition, this case supports geographic expansion and the reduction of operational costs.

11.5.6 Benefits

The punctuality of deliveries has been increased by avoiding the use of roads. Road congestion has been reduced by transferring goods to rail and as consequence greenhouse gas and noise emissions have also been reduced. Regarding economical benefits the practice has achieved a an almost fivefold reduction of societal costs.

11.5.6.1 Economic

- Goods are now transported with almost no damage, which reduces losses.
- The punctuality of deliveries has been improved due to avoiding traffic congestion, so more goods are now available to customers.
- Rail transport is not used by other supermarket chains in Spain, even though the costs are very similar to road transport and although Renfe has introduced "the custom train" (a customised service) to fulfill customer demand.
- In summary, the transport process has become more efficient and Mercadona has achieved an external costs saving of €13.1 million.

11.5.6.2 Environmental

- A CO_2 emissions reduction of 12,000 tonnes per year has been achieved by choosing rail transport.
- Fuel consumption has been cut with less truck use, and there has been a 70% energy consumption saving.

11.5.6.3 Social

- The number of accidents has been reduced due to reducing the number of trucks in use.
- Noise emissions and traffic congestion have also been reduced by unloading trucks at night.

11.5.7 Transferability

11.5.7.1 Transferables

Companies that have a large volume of goods to transport can use this mode of transport as an alternative to road transport. It can be transferred to any country without difficulty for internal transport. The crucial factor is the company's size. Not all companies can involve their suppliers in seeking to change transportation mode in order to reduce costs and increase overall performance for all the agents involved.

Sector transferability is weak in relation to food and perishables.

11.5.7.2 Limitations

It can be used in all countries, but not in international transport across all borders, due to problems with different rail gauges. It is also difficult to use rail for small companies which do not need to transport large volumes of goods. Geography can be a disadvantage where is little potential for good quality rail infrastructure.

11.6 e-Integration and Supplier Evaluation: Siemens – e-Integration

Sonia Guerola-Pérez

Siemens S.A. suffered a lack of communication with its suppliers which gave rise to low productivity and poor service levels.

Siemens S.A. has introduced greater integration of its suppliers through the internet to achieve an integrated supply chain management network. The suppliers' performance is measured on criteria such as on-time delivery, availability of materials, environmental aspects – reducing waste – and their adherence to a code of conduct.

This way the supply chain will become more flexible and competitive, market share will be increased, inventories will be reduced and the service level will be higher than before. These improvements have as a final consequence a reduction in costs.

11.6.1 Problem Description

Due to globalization and the dynamic market where Siemens S.A. works in, the company wanted to achieve good service levels and a high productivity.

The supply chain was accordingly slow to respond to changes (inflexible), and decisions took a long time to make: to sum up, the chain was not as competitive as Siemens wanted.

One of the problems was the lack of communication between the company and its suppliers which was causing delays, misunderstandings, very long lead times, high inventory levels, high costs, etc. Siemens S.A. lacked any kind of measure control and evaluate its suppliers which made it more difficult to take decisions.

The main point for Siemens was to ensure the availability of material in their facilities to increase flexibility and speed in the supply chain, but with the short-term relationships with its suppliers was not easy. In addition suppliers needed visibility in the sourcing process to be more efficient in delivering products.

Then Siemens decided to change the strategy with its suppliers creating longer collaborations through integrating their suppliers in their processes. They realized that they needed an appropriate communication systems between suppliers and Siemens with an evaluation system that guarantees measure the suppliers performance in terms of speed, availability of materials, service level, and so on.

11.6.2 The Solution

11.6.2.1 Creation of a Communication System

Siemens S.A. needs to create a competitive tool for managing the supply chain, increasing its market share and creating a well-known supply chain by moving towards Supply Chain Excellence. That meant that the company should collaborate and integrate with its suppliers, who needed to become customer- and process-oriented.

The first task in achieving these objectives was to develop a system where suppliers and Siemens S.A. communicated via the internet, because communication online is very easy, cheap and an effective way of sharing information.

In this system, the suppliers manage the inventory that Siemens S.A. needs and follow a supplier process evaluation. The degree of difficulty in using the system is not very great, but as it is a new tool, employees need training to use it properly to obtain the best results.

11.6.2.2 Supplier Evaluation

Suppliers are evaluated on four different factors: purchasing, quality, logistics and technology. They are given a fulfillment rate for each of these fields and when the evaluation has been completed, the supplier's performance is marked.

KPIs were set for the evaluation process, such as: on-time delivery of supplies, compliance with requested date of supply and delivery, the delivery capability of suppliers, the quality of supplies, requested date quality, availability of material, first pass yield, delivery reliability and capability, etc. Once the evaluation has been completed, a graphic is displayed in which the supplier's rating is compared against the average of other suppliers.

These tools make the chain become more flexible, because it is possible to choose the supplier that best fulfils a given requirement. Waiting times and inventories are reduced and lead times are improved because it is possible to access information about the availability of various components.

This evaluation is useful not only to assess suppliers, it can also be used as an operational tool to help decide which supplier to choose at any time.

The objective of the graphic is to show the deviations of a given supplier compared to its competitors. In addition, Siemens S.A. requires that its suppliers follow a code of which includes:

- Prohibition of corruption and bribery
- Respect for the human rights of employees
- Prohibition of child labour
- Care for employees' health and safety

11.6.2.3 Sharing Information

With this evaluation, every Siemens S.A. user can access information about a supplier and choose between different suppliers according to the requirements they need to fulfill. This information-sharing was facilitated by contracts between the parties in the supply chain to ensure privacy concerning their business data.

11.6.3 Strategic Implementation and Continuity

As long-term relationships have been established with suitable partners in the inland waterway networks, all the parties involved are committed to developing the solution and ensuring that it will work in the long term.

Due to the reduced environmental impact, the solution is sustainable. Moreover, the solution is in line with Baxter's strategy of the "green supply chain" geared towards the continuous reduction of the carbon footprint of its products.

In addition, the solution is not only less costly than earlier transportation solutions but also and more reliable, paving the way for increased use of inland waterways in the future.

11.6.4 Challenges

Challenge 1: Suppliers were not very keen on making private information about their businesses public.

Challenge 2: Developing a powerful e-system which was easy to use.

Challenge 3: Developing good processes and KPI's useful not only for evaluating the suppliers but also as a daily work tool.

Challenge 4: Training people for the uses of the system was a challenge because workers do not like changing their working procedures.

11.6.5 Lessons Learnt and Success Factors

- Lesson 1: When a communication system is created and the suppliers are evaluated, the chain becomes more flexible and dynamic, inventories are reduced, lead times are improved, benefits rise for all the agents that belong to it, etc.
- Lesson 2: Measurement to control and compare suppliers provides useful information to the members of the chain.
- Lesson 3: Including suppliers in joint activities is profitable for all parties because sharing information and collaborating improves their performance.
- Success factor 1: Siemens S.A. has aligned with its suppliers to be able to develop the system and to integrate it. They have achieved it through contractual agreement with these suppliers.
- Success factor 2: Regarding the processes, the company needed to standardize the suppliers and the procurement.
- Success factor 3: Siemens S.A. created, developed and implemented an information-sharing system to show information about its suppliers.
- Success factor 4: The company needed to create specialists in procurement. It provided training for employees regarding these issues.
- Success factor 5: After improving the process, relationships with suppliers became better and stronger.

11.6.6 Benefits

11.6.6.1 Economic

- Customer service level increases due to reduction of lead time, ordering process time, track and trace order time. It raises product and service availability because of better management of production and deliveries. Customers changed

their ordering habits and could manage better and faster with their suppliers thanks to the website. The process becomes more efficient.
- Inventory holding costs were reduced thanks to a reduction in inventory levels. The system keeps customers informed about their suppliers at any time and with reliable information.
- Decreasing costs and sales growth have made it possible for the company to offer high quality products to its customers at a better price. Integration of the different agents in the supply chain has increased the flexibility of the chain, and has made it able to achieve faster responses.

11.6.6.2 Environmental

- Missions of CO_2 and fuel consumption have been decreased because the number of trucks used has been reduced as a consequence of the improvement in the reliability of deliveries; because fewer mistakes are made, fewer transport journeys need to be carried out.
- The use of paper was reduced by 95% due to the use of the internet.

11.6.6.3 Social

- Because of the Siemens S.A. code of conduct suppliers must ensure the safety of their employees and take care of their health, they must also respect employees' human rights employees and they may not hire child labour. Suppliers must also comply with the prohibition of corruption and bribery. The number of trucks used was reduced and accordingly noise emissions have been lowered.

11.6.7 Transferability

11.6.7.1 Limitations

This project can be implemented in any country and in any type of sector; no limitations are found.

11.7 Optimization of Goods Collection: Mapei – Optimising Goods Collection Cycle Time

Marielle Labrosse

MAPEI, a global leader in chemical products for the building industry, has shown particularly rapid growth over the last 10 years. Increased production at the company's plant at Mediglia (Italy), has created serious truck traffic congestion at and around the plant, with associated environmental and safety impacts. To meet the demands of an increased number of customers and to ensure rapid deliveries, efficient just-in-time logistics processes have been implemented, by optimising the cycle time for the loading of consignments at the factory.

11.7.1 Problem Description

Production at the MAPEI plant in Mediglia is composed mainly of powders, liquids and pastes:

- 80% of products are powders which are packed in bags of 24 kg and transported on pallets, with 50 bags per pallet.
- 20% are liquids and pastes, either in small packages (from 250 ml to 1,000 l) or in bulk. Bulk represents 35% of the liquid products.

The facility in Mediglia was built in 1975. More than 635,000 tonnes of product for the building industry were produced there in 2005, approximately 23% of total production being exported overseas. The surface of the plant is 160,000 m^2, including 53,000 m^2 of covered space. Average production is 2,700 tonnes/day.

Inbound (raw materials) and outbound flows (finished products) are carried by truck, representing 250 trucks/day (average). Sixty percent of these trucks are operated by hauliers who deliver the products to the customers' sites, and 40% are operated by customers who collect their deliveries direct from the factory.

MAPEI's growth has been very significant in the last 10 years. The average daily shipment volume increased from 1,892 tonnes in1999 to 2,670 tonnes in 2007.

Due to the increased production volume and the number of trucks arriving at and leaving the plant, traffic congestion increased around the plant with trucks queuing in the surrounding roads and at the plant's entrance, leading to negative environmental and safety impacts. Moreover, MAPEI wished to increase the quality of the service provided to their customers by reducing truck waiting times and providing additional services to the truck drivers.

The logistics system was therefore re-organised in order to optimise the cycle time for load picks-ups from the factory.

11.7.2 The Solution

In three successive steps, MAPEI invested new funds, restructured its organisation and ironed out any impediments to quick turnaround times at the factory:

1998–99: The plant was reorganised from a logistics point of view, with the creation of new entry points, preventing queues of trucks. The following changes have been implemented:

- Modification of the plant layout
- New facilities for trucks waiting to enter the site
- New shipping offices
- New docking stations, allowing increased loading capacities
- New truck park inside the plant

2004: A new storage tank was built with high-speed loading pumps for bulk liquids, to cycle time.

2005: Logistics processes were modified, with overnight preparation of shipments of pre-packed products.

In addition the environment for truck drivers was improved, with the creation of dedicated washrooms and waiting areas and access to the plant's cafeteria.

Much of MAPEI's business activity is influence by the seasonal factors, since it supplies the construction industry. MAPEI estimates that production volume is 20% higher than the annual average rate in spring and summer, and 20% lower in the autumn and winter.

In agreement with the unions, a seasonal working time scheme was implemented: the employees in the production facilities work four more days per month from 30th April to 30th September (generally working on Sundays), and four fewer days in winter (generally not working on Mondays or Fridays).

These changes have been welcomed as improvements by the employees concerned.

These changes have resulted in:

- Reduction of average loading time, now standing at 78 min. As a comparison, in the MAPEI plant in Rome (producing similar products) average loading time is 130 min.
- Reduction of long waiting times (trucks no longer queue outside the plant).

11.7.3 Challenges

Challenge 1: In order to ensure rapid delivery of the goods ordered by more than 40,000 customers around the world, the biggest challenge was the implementation of an efficient logistics system embodying a just-in-time approach.

Challenge 2: The main challenge was the ability of the logistics process to control and reduce any bottlenecks between the time customers' trucks arrive at and leave the site with the right load, cutting any unnecessary waiting time, particularly outside the plant.

Challenge 3: The optimisation of the cycle time involved several factors across the supply chain and production processes: a new plant layout with docking stations, revised production capacity, the timing of the preparation of shipments, and an improved environment for truck drivers.

11.7.4 Lessons Learnt and Success Factors

- MAPEI's overall strategy was to focus on the quality of services delivered to the customer, and on the environmental and safety issues raised by trucks queuing.
- The cycle time of trucks within the plant resulted from different logistics processes which could only be improved over time. This is why this reorganisation was carried out in three successive steps beginning in 1998.
- The individual changes implemented by MAPEI are relatively "standard" in terms of practices and improvements. The innovation relies on the overall control that MAPEI can exercise over the cycle time of every truck within the plant.

11.7.5 Benefits

11.7.5.1 Economic

- Average loading time has been reduced from 130 to 78 min (−40%).
- The quality of service delivered by MAPEI has increased, with trucks no longer queuing outside the plant.
- This new organization has improved the level of customer service offered with the implementation of additional services offered to truck drivers (bathroom, cafeteria, etc). It also improved the availability of the service by being able to cope with the seasonal demand variations of the construction business.

11.7.5.2 Environmental

- This new organisation has drastically reduced the number of trucks queuing outside the plant, which were creating bottlenecks and traffic jams, and therefore has contributing to the reduction of trucks' emissions.

11.7.5.3 Social

- The new logistics organization of the plant and associated services has provided better working conditions for the truck drivers.
- Seasonal management of working time for the MAPEI employees has been implemented in consultation with the union and has been seen as positive change by the employees.

11.7.6 *Transferability*

11.7.6.1 Transferables

Industry: Such a logistics reorganization could be accomplished in other types of manufacturing industry, wherever cycle time and the quality of services are key issues for customers.

Geography: The reorganization is not related to a specific geography. The main factor here is the availability of space, in particular for the creation of shipping offices, new entrances, etc. The plant of Mediglia is now a reference model for the other production facilities of the MAPEI Group.

11.7.6.2 Limitations

Company size: The large reorganization (implying significant investment) restricts this practice to relatively large companies.

11.8 Railway Transport: Carpathia Express – Transport Collaboration

Miroslav Rumler

The productivity of rail transport has been steadily declining in recent years. This fact, together with a desire to develop a new innovative product for existing as well as potential customers, has lead a team of logistics professionals to design a brand new commercial logistics project. It builds on the close cooperation of these three traditionally competing freight forwarders and three national rail transport providers with the ultimate aim of maximizing the resource utilization of all parties involved and at the same time increasing the productivity and competitiveness of rail transport.

11.8.1 *Problem Description*

The development of transportation during the twentieth century and in recent years in the Czech Republic has mostly favoured road transport, while rail transport has been left behind. In summary:

- Road transport has grown 2.5× (by 1997)
- Freight volumes in rail transport have declined by 50%

- Freight volumes steadily declined until 2002
- A pessimistic estimate of the annual revenues of the Czech national rail freight provider for the three years from 2002 suggested further decline of some 17%
- The ratio of rail/road transport volume is 24.5% (rail) to 71.9% (road)
- As market development brings ever larger percentages of small and lightweight shipments into total freight volumes, the competitive advantage of rail transport has been declining substantially
- Some local rail tracks have been closed
- Large number of traditional manufacturers have either shifted their operations to other markets or closed their railway sidings
- Government support of rail transport has recently been very weak

This background, together with the recent growth of environmental concerns and a desire to offer progressive and innovative services, resulted in the launch of the CARPATHIA EXPRESS project and the formation of the Carpathia Group, an independent train operator owned by three established freight forwarders from the Czech Republic and Germany in cooperation with three national railroad operators.

11.8.2 The Solution

The name "CARPATHIA EXPRESS" clearly illustrates the route on which the Group's shipments are transported. The project partners were not chosen randomly – they comprise forwarding agents and railroad companies. It is forwarding agents, as well as logistics providers, who decide on which route and which transportation mode will be used in individual cases.

Railroad companies are usually in charge of the entire infrastructure and a full range of rail coaches; Moreover, they have considerable expertise in various international transport technologies. Traditional competitors – Argo Bohemia s.r.o., Spedi-Trans Praha, s.r.o. a Raabersped Speditionsgesellschaft m. b. H. – each market leaders in their own logistics markets – decided to override their own corporate interests for the sake of this project in favour of a joint effort. Each project partner took advantage of its knowledge of and professional expertise in the flow of shipments from the CEE region to Balkan markets. All researched carefully the pros and cons of individual transport modes in terms of their capacity, quality, speed, etc. on one hand, and the requirements of customers on the other.

The following results have been achieved:

- A working team emerged leading to the establishment of the Carpathia Group
- Business policy was set and a reservation system launched
- Operational rules were defined for train technology
- A decision was made that the CARPATHIA EXPRESS logistics train will be available for public use.

The CARPATHIA EXPRESS logistic train is a fast integrated commercial connection designed for the movement of truck load shipments and containers from the CEE region to South Eastern Europe. CARPATHIA EXPRESS is open to all interested parties who may wish to use it for their shipments via Carpathia Group. Carpathia Group acts as an independent train operator.

CARPATHIA EXPRESS freight trains are open for all types of truck load shipments from domestic and international shippers with the exception of the following:

- Military transport
- Private empty rail cars
- INTERFRIGO-governed shipments
- Oversized shipments
- Livestock

The first shunting station for CARPATHIA trains is Havlickuv Brod. Other shunting stations are Brno-Malomerice, and Breclav. The train's terminal destination is either the city of Curtici where shipments are sorted and distributed to their final destinations (Route length, 715 km, journey time 27 h) or alternatively the city of Episcopia Biharhere, where shipments are sorted and distributed to their final destinations in northern Romania and Moldavia (722 km, journey time less than 28 h).

The technology of CARPATHIA EXPRESS allows it to accept consignments from Western Europe via Czech-German and Czech-Poland border crossings.

It is vital for the project that the service can meet JIT logistics technology requirements. CARPATHIA EXPRESS also supports standard hub-and-spoke logistics patterns, being regular, fast, spacious, and providing reliable connections, which allow for the creation and use of a logistics centre for collection and delivery of shipments, and for their re-distribution.

11.8.3 Challenges

Challenge 1: Pooling the individual customers of the project partners into services subsequently provided by the Carpathia Group

Challenge 2: Forming a common sales strategy

Challenge 3: Increasing productivity and countering negative trends in the transport industry, despite the overall decline of rail freight volumes (on a given route)

Challenge 4: Radical reduction of contemporary standard transport journey times

Challenge 5: Cut overall costs in a way that enables operators to cut the prices charged to their customers substantially

Challenge 6: Increase in competitiveness

Challenge 7: Ensuring transport and pricing flexibility while adhering to fair-play principles

11.8.4 Lessons Learnt and Success Factors

- The cooperation of formerly competing companies yields substantially reduced costs for all partners involved, and ultimately, also reduces final transport costs for most customers
- A well-defined sales policy, together with the right technology, can yield a remarkable increase in rail transport volumes regardless of the overall stagnation of the rail transport market in Europe
- When offered regular, reliable and fast services at competitive prices, customers are willing to shift their shipments from road to more environmentally-friendly modes
- Partnership of industry competitors with similar transport streams can generate a critical mass of individual shipments which can hardly be generated on so large a scale by a single entity

11.8.5 Benefits

- Dramatic reduction in journey time
- Faster turnover and improved loading efficiency of rail coaches
- Cost reduction which ultimately leads to the reduction of transport costs for individual customers
- Increased competitiveness of rail transport between Czech (and the broader CEE region) and Romania (and the surrounding markets)
- The combination of shipments generated by individual partners helps reach critical mass

It is very important that the technology applied meets the requirements of customers who prefer JIT and/ or Hub-and-Spoke scenarios.

11.8.5.1 Economic

- Average 66% saving on journey times compared to other transport modes
- High quality "one-stop-shop"-type of service for project customers including check-in, space reservation, despatch operations, etc.
- The product developed product is widely available and beneficial not only for local customers, but also for customers located in western and northern Europe
- The original customer portfolio during initial 6 months of commercial operations was extended by 400%

11.8.5.2 Environmental

- CO_2 emission reduction as a result of the transfer of a remarkable quantity of shipments from road to rail
- As a result of the centralization of shunting operations to just a few railroad railheads, the noise emissions caused by rail tractor units (engines) have decreased substantially along the entire CARPATHIA EXPRESS project route(s)

11.8.5.3 Social

- N/A

11.8.6 Transferability

11.8.6.1 Transferables

Implementation of similar schemes elsewhere is feasible and is not limited to any particular railroad company, forwarder, or location.

11.8.6.2 Limitations

The only limitation is the presence of a real desire shared by several entities to share their know-how and customer portfolios in order to enhance their business opportunities, with the ultimate aim of increase overall business volumes and individual revenues.

11.9 Consolidation by DC: Hamé – Distribution Centre

Miroslav Rumler

Hamé a.s., the leading Czech food manufacturer, has experienced rapid growth of its operations in the last few years not only in the Czech Republic, but also in other European countries – Slovakia, Poland, Hungary, Ukraine, Moldavia, Romania and Russia. The vast increase in flows of goods and the strict requirements applying to the majority of Hamé production lines created an urgent need for the construction of a new customized distribution facility with European coverage. The chosen combination of the solution concept and the technologies used makes the approach unique and ensures that it meets all the requirements defined in the preliminary stage of the project. As a result, Hamé has cut its logistics costs, reduced negative environmental effects by making substantial cuts in total kilometers driven and reduced the damaged goods and error rates as well as the amount of time the goods stay in the DC; the solution is also characterized by increased utilization of technologies, improved labor safety, higher productivity, increased service speed and better service accuracy.

11.9.1 Problem Description

The original situation was criticized for system inconsistency, the low level of cohesion (flow warehousing, limited floor footage, more than one forwarding department, external buffer warehouses for peak volumes and multiple trips between individual sites), low transmission quality and delays in information flows, the handling of goods and paperwork in depots, manual processing of documents, the high rate of damages, the high error rate and additional costs due to the transport of goods from plants to depots, buffer warehouses, and between individual depots. Problems were also identified in the following areas:

- Complicated distribution system
- Long response time to customer orders with limited options for the consolidation of final deliveries.
- High warehousing and transport costs.
- Traditional warehousing technology with a high proportion of manual work.

The company has defined the parameters of the new facility in depth; using modeling and suitable simulation techniques, the most appropriate warehousing and handling technologies have been designed in cooperation with various suppliers.

The goal of the new set-up is provide a solution which simplifies day-to-day operations, speeds up order response time and optimizes utilization of both warehousing technologies and transportation capacity. If the company had not decided to pursue these ambitious goals, its market competitiveness would have been severely threatened in years to come.

11.9.2 The Solution

The massive increase in flows of goods and the strict requirements for the majority of Hamé production lines have recently resulted in the construction of a new distribution center tailored to the needs of the company.

Parallel to this, Hamé launched a special project with the ultimate aim of developing a tailored technological solution based on a unique combination of the most state-of-the-art technologies offered by the world's leading manufacturers of warehousing equipment.

The warehousing and dispatch system used in the European distribution center of Hamé comprises standard line shelves, BT Radioshuttle deep-reach warehousing and a Jungheinrich Drive-In_System (DIS), substantially increasing the efficiency with which the existing warehousing space is used. The combination of DIS and the central dispatch tunnel is unique in our environment; this solution is used for the dispatch of combined pallets with high turnover rates in order to radically speed up the entire process.

Jungheinrich Drive-In-System with dispatch tunnel with sloping roller conveyor featuring hi-density flow shelf system with a capacity of over 2,880 pallet positions, continuous pallet dispatch in the tunnel, controlled shift of pallets in flow channels by satellite trolleys.

BT Radioshuttle is also a hi-density flow shelf system offering a capacity of 3,920 pallet positions, continuous pallet dispatch in the tunnel, a depth of 10 pallets for full truck loading, controlled shift of pallets in flow channels by satellite trolleys WiFi technology with remote (wireless) terminals and bar coding.

Advanced handling technology structured and designed on the basis of mathematic modeling. Brand new DC building meeting all modern standards with main warehousing area, inbound section and outbound section.

As a result, Hamé successfully reduced logistics costs, negative environmental effects and the time the goods stay in the DC; it increased utilization of technology, work safety, productivity, service speed and service accuracy.

In addition, the chosen solution sets new standards for the internal flow of information with the highest possible level of its security. State-of-the-art technology is driving the demand for higher staff qualification – and this will result in the creation of a tailored corporate training system for company employees on all levels. All this positively influences the corporate culture, reflecting the changes occurring at all levels of the company.

11.9.3 Challenges

- Improvement of warehouse KPIs
- Redesign of company processes
- Online communication between the warehouse management system and company IS
- Reduction in external "buffer" warehousing operations
- Merging of 5 dispersed dispatch centers into just 1
- Capacity increase and stock optimization
- Use of latest handling technologies (automatic identification, WiFi, modern remote terminals, etc.)
- Staff reduction
- Focus on environmental issues

- Reduction of handling times (resulting in lower energy consumption)
- Reduction in the frequency of errors and damage

11.9.4 Lessons Learnt and Success Factors

- Implementation of a new WMS delivered instant results by highlighting numerous redundant functions, inefficiency and other examples of poorly planned utilization of systems which – despite the well-meant efforts of personnel on all management levels – were no more than the logical result of obsolete technology.
- New people, new layout, new building, new technology – all these factors created brand new challenges for the management as well as for all staff.
- The initial phases, especially in terms of reaching consensus on the necessity for change in order to increase overall efficiency, were the toughest ones in the entire project. The reason for this was that some of the people who had been "emotionally involved" in driving the huge growth of the company were forced to make more "rational" decisions.

The solution is unique and innovative with regard to the:

- Unprecedented combination (in the Czech Republic) of two state-of-the-art warehouse technologies and their integration to form one common warehouse system.
- Implementation of an advanced warehouse management system which takes account of both the features of the deployed technology and the requirements of company logistic processes.
- Highly beneficial ratio between the necessary investment and the benefits achieved.

11.9.5 Benefits

Flexibility, the quality of all logistic operations, accuracy, labor productivity, staff qualification, work safety and profit margins have all increased since the project was launched; at the same time, system failures, discrepancies, accidents, injuries, warehousing costs, transportation costs, transportation needs and total personnel costs have been reduced.

11.9.5.1 Economic

- The chosen solution combines economic benefits of two kinds: firstly, it centralizes existing operations in a single facility, which substantially reduces logistics costs (while achieving better results with fewer negative overall effects). At the same time, it also ensures improved utilization of state-of-the-art technology by achieving a critical mass of items handled.

11.9.5.2 Environmental

- Significant reduction in the number of unnecessary truck trips to customers.

11.9.5.3 Social

- The new logistics organization of the plant and associated services has provided better working conditions for the truck drivers.

11.9.6 Transferability

11.9.6.1 Transferables

The selected solution can be applied anywhere regardless of industrial, geographical, or company specifics – the only limitation being the partly customized adaptation of the warehouse management system to specific company processes and the in-house information system. Warehousing and handling techniques would have to be tailored to the actual product lines or product ranges.

11.9.6.2 Limitations

Low volumes of stored and handled products; small company size which would not justify substantial financial investment.

11.10 Transport Marketplace: Sharp – Sharp's Collaboration with the Green Transport Marketplace, SmartWay

Mark Servidio and Balkan Cetinkaya

Sharp Electronics Corporation is a global electronics manufacturing company based in Osaka, Japan. One of the company's core values/commitments is reducing environmental impacts including minimizing greenhouse gases. Sharp in the USA wanted to explore how they could contribute in the transportation area. Sharp decided to join a transport partnership – called SmartWay – which was implemented by the US governmental organisation EPA. A partnership between shippers and carriers who committed to operating green transport practices, which also lead to increased revenue and cost savings.

11.10.1 Problem Description

The growing concern and worldwide mandate to improve the environment caused Sharp Electronics Corporation to recognize the problems and issues of greenhouse gas emissions and impacts on air quality. Sharp Electronics established a goal to reduce their impact on the environment. According to the EPA (Environmental Protection Agency), in the United States, trucks travel over 200 billion miles a year using 55 billion gallons of fuel. This equates to 27% of all US oil consumption and 13% of all carbon emissions. Trucking also contributes to 56% of the NOx emissions and 32% of PM emissions.

The issue is while there is a lot of work being done on recycling paper and plastics, and on manufacturing activities related to the environment, there were only limited activities in the logistics area. Sharp's goal was to get involved in the area of transportation, based upon the above statistics. The target was to implement four key initiatives dealing with logistics and the environment.

One of these was for Sharp Electronics USA to partner with the EPA to achieve reductions in CO_2, NOx, and PM within the United States under a voluntary green transport marketplace program called the SmartWay Transport Partnership.

By doing so Sharp Electronics would met the Corporate Balanced Scorecard on CO_2 emission reductions with a goal of the company and its products having zero emissions by 2012.

11.10.2 The Solution

Sharp's first step was to educate the corporation on the SmartWay Transport Partnership.

SmartWay Transport is a voluntary partnership between various freight industry sectors and the EPA that establishes incentives for fuel efficiency improvements and greenhouse gas emissions reductions. By 2012, this initiative aims to reduce between 33 and 66 million metric tons of carbon dioxide (CO_2) emissions and up to 200,000 tons of nitrogen oxide (NOx) emissions per year. At the same time, the initiative aims to achieve fuel savings of up to 150 million barrels of oil annually.

There are three primary components of the programme: creating partnerships, reducing all unnecessary engine idling, and increasing the efficiency and use of rail and intermodal operations.

Sharp focuses on utilizing SmartWay carriers for at least 50% of its shipments within 3 years. Carriers commit to implement actions which will reduce emissions, such no truck idling, low sulphur diesel, and installing diesel scrubbers. The core team worked with Sharp-approved carriers to understand the partnership and established a plan that included the following:

- Communication of a "No-Idling" policy for truckers working with Sharp's Logistics Centers
- Promoting carriers that used low emission equipment and higher fuel standards
- Increased intermodal shipping
- Driver Comfort stations to promote less idling while waiting for their appointments
- Increased commitment to use SmartWay Transport Partnership carriers at least 50% of the time within 2 years

The solution to avoid having to find new carriers was to get current carriers to become SmartWay members. Sharp was very successful in getting 34 of 35 carriers signed up. From the carrier perspective the EPA provided tools for the carriers to improve their processes.

The SmartWay Transport program works with states, banks, and other organizations to develop innovative financing options that help partners purchase devices that save fuel and reduce emissions, and shows companies how they can reduce emissions and save money by using the SmartWay Technology Package Savings Calculator.

Lastly the big benefit of SmartWay was that Sharp used it as a platform to become more involved in other environmental projects. While some of these are small they have got people involved:

1. Installing motion detectors at a Logistics Center, saving electricity
2. When travelling, only renting hybrid cars
3. Using recycled paper in faxes, copiers, and printers
4. Printing double sided copies
5. Giving CFL lights to employees
6. Test using corrugated honeycomb recyclable pallets

The project manager found that the level of interest in his department increased, as the management saw environmental issues were not just the job of the Environmental Manager and not just a matter of recycling paper.

Sharp has a Corporate Social Responsibility annual report and a corporatewide environmental section of the company's strategy. The core values include environmental awareness. This programme is now part of Sharp's ISO 14001 programme.

11.10.2.1 Implementation

Number of SmartWay Carriers of Sharp USA

- 2004–05 – 24 Out Of 85 Total Carriers (28%)
- 2005–06 – 39 Out Of 85 Total Carriers (46%)
- 2006–07 – 72 Out Of 80 Total Carriers (90%)

- 2007–08 – 39 Out Of 47 Total Carriers (83%)
- 2008–09 – 39 Out Of 40 Total Carriers (98%)
- 2009–10 – 32 Out Of 32 Total Carriers (100%)

11.10.2.2 Tons Shipped with SmartWay Carriers of Sharp USA

- 2004–05 – 41,808 Out Of 127,841 Total Tons (33%)
- 2005–06 – 81,499 Out Of 102,808 Total Tons (79%)
- 2006–07 – 113,546 Out Of 115,468 Total Tons (98%)
- 2007–08 – 104,415 Out Of 108,313 Total Tons (97%)
- 2008–09 – 107,265 Out Of 107,269 Total Tons (100%)
- 2009–10 – 73,836 Out Of 73,836 Total Tons (100%)

11.10.3 Challenges

- Challenge 1: Reduction of greenhouse gas emissions with minimal to no impact on the quality of service that Sharp provides to its customers in the areas of transportation and logistics.
- Challenge 2: Whether going to new carriers as part of the SmartWay program would cause service issues was a challenge, as getting current carriers to sign up was critical to achieving zero service changes.
- Challenge 3: Find a quality solution without impacting the costs of transportation and logistics services.
- Challenge 4: The modernisation of the fleet/carriers in US was effected due to various laws taking effect, not by SmartWay.
- Challenge 5: Convince Sharp-approved carriers to join us in becoming an EPA SmartWay partner or risk having new carriers handle business, increasing risk.

11.10.4 Lessons Learnt and Success Factors

- Being one of the programme's first partners and pioneers has given Sharp credibility in the corporate environment and influence in this field.
- The EPA SmartWay Transport Partnership provides resources and tools that are no cost to the shipper.
- From the carrier perspective the EPA provided tools to the carriers to improve their processes. The SmartWay Transport Program works with states, banks and other organizations to develop innovative financing options that help partners purchase devices that save fuel and reduce emissions and show companies how they can reduce emissions and save money by using the SmartWay Technology Package Savings Calculator.

- Increase awareness of environmental issues. Having the team understand SmartWay has caused them to think of other ways to become involved.
- Commitment from carriers to the SmartWay Transport Partnership was a key milestone in order to make the partnership work. Getting current carriers to join the programme allowed Sharp to continue using the same carriers, thus causing no service disruptions.
- The cost was minimal to Sharp Electronics while making a substantial impact to the world we live in. Overall, according to reports from the carrier, the investments they made were offset by the fuel reduction they achieved. So to them it was cost neutral with business gain.
- Peer awareness helped convince Panasonic & Sony to join the SmartWay program.

11.10.5 Benefits

11.10.5.1 Environmental

Reduced diesel fuel consumption. Reduction in electricity from motion sensors in California Logistics Center. From October 2004 to September 2007 Sharp has calculated that by shipping with SmartWay carriers they have executed the following improvements:

- Reduced CO_2 emissions by 1,383 tons, NOx by 26.5 tons, particulate matter by 1.1 tons.
- Increased percentage of rail shipments from 7% to 12% of the annual tonnage.
- Implemented "No-Idling" policy at Logistics Centers.
- Increased the percentage of Sharp's SmartWay carrier usage from 33% tons in the first year to 97% by the 4th year.

Sharp also took smaller steps by using recycled paper in copiers, renting hybrid cars when travelling and carrying out preventive maintenance on forklift trucks.

11.10.5.2 Social

Since joining SmartWay, Sharp has been recognized by consumers, partners, and internal management as being a leader in the promotion of Corporate Social Responsibility. A sense of community has been established revolving around the stewardship of the environment.

As part of Sharp Electronics Corporate Social Responsibility programme this fits into the Corporate Strategy. For those that know of SmartWay it has made Sharp look like a better corporate citizen.

11.10.6 *Transferability*

11.10.6.1 Transferables

This case is relatively independent of country and sector characteristics.

11.10.6.2 Limitations

Needs to be adapted by an independent body, ideally for the whole of Europe, to get it to work. The development of such a platform by a SME is very limited.

11.11 Transport Exchange: British Telecom – Improving Efficiencies Within a Scheduled Trunking Service

Lyall Cresswell supported by Peter Franke

This case study shows the contribution which online freight exchanges can make towards long term collaborative logistics partnerships. These can lead to a significant reduction of freight traffic and improved efficiencies for all stakeholders. Based on the example of a major shipper/service provider, opportunities, challenges and mechanisms in using a freight exchange are presented.

11.11.1 *Problem Description*

In common with many businesses, BT Group plc has a "scheduled" time definite trunking network to service its UK distribution depots. This haulage is carried out by own vehicles, mainly using double-deck 13.6 m curtain-side trailers.

As an own-account operator, BT seeks to maximise cost recovery in this network by charging their suppliers for the uplift of goods which have been purchased on "ex works" terms (so-called "factory-gate pricing").

The BT Supply Chain organisation required a solution to maximise load utilisation on certain lanes and meet their CSR obligations to minimise carbon footprint within their transport division. Therefore, they looked for an effective way to advertise their scheduled vehicle movements to the wider logistics market with a view to sourcing suitable partners for sharing capacity.

BT's Corporate Social Responsibility programme is driven by:

- The need for sustainable economic growth
- The need for wider inclusion of all sections of society

- The need to tackle climate change
- This is in line with bestLog's three dimensions of sustainable logistics practice.
- BT regularly publishes KPI and targets for its CSR programme. These targets relate to all areas of BT's business including transport.

11.11.2 The Solution

Haulage Exchange is the UK's pre-eminent freight exchange for the road transport industry. Over 2,000 transport companies use the exchange to share available loads and empty vehicle movements, thereby gaining additional revenues.

The exchange is totally impartial and operates on a subscription basis – this ensures that users are comfortable with entrusting the exchange with their sensitive commercial information.

The majority of movements are in excess of 100 km, and are generally FTL although there is a growing volume of part-load offers.

In excess of 0.5 million movements per annum are traded between the members.

In summer 2008, the decision was taken to develop a new section of the exchange: "*Regular Runs*". Freight exchanges have traditionally focused on the 'spot' or ad-hoc market – however, the UK logistics market is heavily driven by contract logistics serving sophisticated distribution networks, many of them operating on a JIT basis.

"*Regular Runs*" enables professional operators such as BT plc to advertise scheduled vehicle movements to a wide audience of transport professionals, and thereby gain a number of benefits.

The process works like this:

- Clients such as BT publish their regular runs onto the Haulage Exchange website, details include:
- Start/finish locations
- Departure days
- Departure and arrival times
- Vehicle type, space availability
- Suitable freight
- Load weight available

  The system accepts multiple runs per day, each with their own details if required.
- The runs are viewable by visitors to the website and members of the exchange.
- An enquiry is made online through the journey listing – at this stage the identity of the journey owner is not disclosed.
- If the owner is satisfied with the enquiry, they validate the application and begin contractual negotiations – these can be recorded in the system. As part of the validation process, the company making the enquiry accepts the Terms and Conditions of the journey owner. Negotiations over rates and contract details are entirely between the two parties, the exchange is totally neutral in this area.

- Once the Regular Run is published it attracts enquiries from many parties. The journey owner is able to create a distribution list for each Regular Run – this enables them to update the daily status of each journey without further manual activity.

From this, it is readily apparent that there are many benefits for companies who choose this solution, including:

- Improved vehicle utilisation – and associated profitability
- Enhanced communication with new trading partners
- Streamlined communications - both internally and externally
- Reduced Carbon Footprint through a simple and effective collaborative mechanism

11.11.3 Challenges

Challenge 1: To persuade transportation and logistics companies to participate in the setup, both at the deep sea port area and the inland container terminal. It was necessary to establish services in the deep sea ports complete with the option of loading on river boats or barges.

Challenge 2: To ensure that transportation costs would not exceed the previous cost of road transportation.

Challenge 3: To ensure that total transit time would not increase with the new solution.

Challenge 4: To find a service provider to set up a stock yard for this solution.

11.11.4 Lessons Learnt and Success Factors

Effective collaborative logistics depends upon several factors:

- A willingness to collaborate with other trading partners, and share volumes
- Reasonable flexibility from both parties – it's unlikely that there will be a 100% perfect match with an organisation's existing runs
- A willingness to embrace new working practices

We can conclude that:

- For small and bigger operators and forwarders alike, freight exchanges are an excellent means to improve utilisation of their fleet and thereby increase profitability
- Freight exchanges hence contribute to the reduction of traffic volumes and the sustainability of logistics
- Transaction cost can be minimised by implementing direct interface with the companies own software systems if necessary

11.11.5 Benefits

11.11.5.1 Economic

- Cost Reduction
- Higher turnover

11.11.5.2 Environmental

- Improved utilisation of trucks, lower emissions

11.11.5.3 Social

- Over a period of time, reduction of traffic as overlapping runs are shared

11.11.6 Transferability

11.11.6.1 Transferables

The solution is transferable in all countries irrespective of company size – the only requirement is that the runs should be of a scheduled nature and that the "owner" of the run should be open to collaborative working and be reasonably flexible where necessary to accommodate longer term changes.

11.12 Influence of Product Design on Warehousing and Transport Efficiency: Bosch Siemens Haushaltsgeraete – Influence of Product Design on Warehousing and Transport Efficiency

Barbara Ocicka under supervision of Krzysztof Rutkowski

Due to copyright issues this case is exclusively available online at http://www.bestlog.org. The case can be found in the "knowledge Base" section at the website.

Chapter 12
Enablers and Support Cases

12.1 Telematics: Sieber – Telematics at SMEs

Christoph Tyssen

Rising fuel expenses are the driver for investments in fuel-saving technologies – even for an SME. Sieber – a Logistics Service Provider – uses telematics solutions in order to reduce the fuel consumption of its truck fleet by monitoring the driving style of its staff. The company engages the collaboration of the drivers in the improvement process. As the drivers were truly convinced of the programme, this project resulted in a nearly competitive contest of the drivers to be the driver with the fewest wearout and lowest fuel consumption.

12.1.1 Problem Description

This example of good practice should be of special interest to SMEs in the road transport business, as it provides solutions for cost reduction and performance improvement.

Rising fuel expenses are a cost factor of increasing importance in the transport business. Furthermore, customers increasingly expect frequent status reports on their shipments, particularly in digital formats.

As the price for transport services is dictated by the market and cannot be influenced by SME suppliers, their main sphere of influence is on their own costs.

Objective data was missing to permit the measurement of trip mileage, fuel consumption and other cost drivers. Using telematics generates the necessary data with ease. This was the reason for Sieber to decide to monitor its trucks with telematic systems. The company expected to reduce overall fuel consumption as well as to continuously monitor wear and tear on truck components.

C. Tyssen et al., *Sustainable Supply Chain Management*,
DOI 10.1007/978-3-642-12023-7_12,

High investment costs used to scare off SMEs from purchasing telematics systems. SMEs are very careful to avoid mistaken investments. If telematics solutions do not pay off through reduced fuel and maintenance costs they represent a poor investment. Drivers may feel controlled and mistrusted if their every move is monitored by telematics surveillance tools. This could result in decreased motivation, and might even lead to increased turnover of driving personnel.

12.1.2 The Solution

These arguments did not stop Sieber from introducing telematic systems in their truck fleet. The decision was not only supported by the expected cost savings, but also through the firm's intention to improve the negative image of road transport in Europe.

Sieber therefore decided to implement the product of a reputable telematics service provider. Drivers and Sieber's repair shop staff needed to be integrated within the new system. The telematics tool provides monitoring of the trucks and analysis of truck performance, and allows direct communication between the despatch department and each driver to enhance order processing and optimise route planning.

12.1.2.1 Dare to Invest

Road transport has the reputation of being a "climate killer" and environmental polluter (ignoring the fact that road transport is indispensable in Europe). Sieber wanted to improve its reputation and decided to invest in technology to save fuel. Despite the aforementioned risks, Sieber therefore implemented telematics in its trucks. To receive the desired results they invested in the system of an established telematics provider and installed it into their trucks.

12.1.2.2 Integrate the Employees

In order to avoid conflicts, employees were integrated into the process of telematics implementation at an early stage. This reduced possible negative effects to a minimum, and even led to cooperative teamwork during the installation phase. Drivers had the chance to get to know the new system and dispel fears of being monitored.

Rather they came to understand how the system could help them perform better. The involvement of all types of personnel and their interaction increased communication and made the common goal of the telematics system clear to everyone. The drivers even felt motivated and challenged to achieve the best possible fuel reduction figures.

Civilised interaction between management, instructors and drivers with the aim of achieving the common goal of fuel consumption reduction can be seen as one of the most important success factors. It helps Sieber remain competitive and every driver now feels that he can really be part of the project, and support it.

As Sieber is aware of the negative image of road transport services, it is interested in improving this perception. Further, fuel cost savings have a direct positive effect on overall operating costs. The savings and performance advantages save from 1.5 to twice the amount spent on the implementation and maintenance of the telematics solution.

12.1.3 Challenges

Challenge 1: Selection of adequate telematics system provider: every SME has to evaluate a telematics provider appropriate for its individual needs. It is especially important that the provider and the user SME communicate on the same level.

Challenge 2: Tailoring the system to the company's needs: as every SME has its own needs and requires customized telematics it is important to have a supplier capable of adapting to these requirements (concerning the hardware and software required and the system interfaces).

Challenge 3: Implementing the new system with regard to operational efficiency and countering employees' fears: As employees are often scared of changes, and especially of the monitoring tools provided by telematics, it is crucial to communicate the rationale for the solution effectively.

Challenge 4: Data handling and analysis, and employee education: In order to use the system efficiently it is important to provide training for the affected employees (concerning the use of the data and associated business improvement activities).

12.1.4 Lessons Learnt and Success Factors

- The technology allows better service levels and fewer accidents if used correctly.
- Investing in technology to reduce fuel consumption is worth the expenditure.
- A collaborative climate between the telematics provider and the LSP is elementary.
- The telematics provider should be capable of understanding SMEs' concerns, while the SME needs to be open to new technologies.
- As an LSP-SME, choose the right product for your needs. As a telematics provider advise your customer wisely and only offer the systems he needs and is capable of using.
- Train your employees in new technologies. Personnel have to buy into the rationale of telematics and have to be capable of handling the telematics system.

- IT implementation, especially with telematics, increases the transparency of drivers' behaviour. Therefore early and intensive involvement of drivers is the key to success.
- Appoint someone in charge of providing feedback to the drivers in order to achieve improved driving performance. The technology only provides the tools; people have to make the difference and understand and execute the suggested actions and improvement measures for sustainable driving.
- Implement a culture of continuous learning. Every employee must understand that he\she must contribute to cost savings and environmental improvements.

12.1.5 Benefits

12.1.5.1 Economic

Telematics monitors the driver's driving behaviour. Conversation between workshop personnel and the drivers helps to improve driving styles. Sieber drivers have saved at least 2 L per 100 km. The savings achieved, from reduced fuel consumption and reduced component wear, e.g. on brakes, are 1.5–2 times the cost of system implementation and maintenance.

12.1.5.2 Environmental

Continual monitoring allows Sieber to adopt optimized truck service intervals. This leads to less pollution and consumption of natural resources as the trucks achieve the same performance as before but with reduced consumption of replacement parts. Besides saving fuel through optimized driving techniques, telematics technology also helps to optimise route planning, leading to shorter journeys, hence to less fuel consumption.

12.1.5.3 Social

The truck-drivers are more relaxed, and can thereby offer a friendlier service. Sieber furthermore has the opportunity to inform customers of possible delays at an early stage. Employees feel better integrated with the company's objectives, and more important to their achievement. They feel themselves to be important elements in the success of the company.

As the cost structure of Sieber is now improved and drivers are less easily replaced due to their increased knowledge, they can concentrate on their job and need not worry about their job security. Driver loyalty has accordingly risen to a much higher level.

12.1.6 Transferability

12.1.6.1 Transferables

Telematics for road transport offers great cost benefits while reducing negative environmental effects. The costs of the installation of telematics tools is quickly repaid in fuel savings. This makes the solution applicable to all types of road haulage in all countries. Company size does not influence the transferability of this good practice within the road transport sector.

12.1.6.2 Limitations

Telematics nowadays is mostly applicable to road transport, with limited transferability to other transport sectors. Company size only limits transferability in terms of the financial resources available for the initial investments.

12.2 IT Integration: Port of Valencia – IT Integration System

Sonia Guerola Pérez

There is a general lack of standardization with regard to the systems used by agents. Furthermore, the volume of container traffic has grown considerably, and this has resulted in an increase in the volume of logistics services and port operations. In order to speed up processes and operations, the port of Valencia decided to create a new IT system to coordinate all agents operating within the port. valenciaportpcs.net is a web service platform which was created to coordinate all the agents involved in maritime logistics processes. This IT platform integrates the following:

1. The EDI services provided by the Port Authority for handling ports of call and dangerous goods.
2. The services offered by the SIC (information and communication system) for handling haulage, manifests, liaison with terminals and customs information.
3. The maritime services provided by carriers: shipping instructions, bookings and schedules.

In short, it offers a range of options that are available to all agents involved in port logistics.

12.2.1 Problem Description

Relationships between the different agents in the harbor are sometimes complicated. The players involved in the process are in most cases ship and cargo agents and operators (forwarding agents, shipping agents, carriers, etc). There is a general lack of standardization in the area of reports, IT systems and the procedures these agents use. Furthermore, processes are inefficient and sometimes manual, the visibility along the supply chain is heavily reduced and there is no integration with back-office systems.

On the other hand, the constant growth of container traffic at the port of Valencia has resulted in a parallel increase in the volume of container-related services and in particular land transport.

This changing situation overtaxed the previous procedures and practices and revealed how inadequate they were. This led to the realization that these practices should be replaced by other methods in which technological solutions play a key role in identifying potential areas for improvement. These new methods should cater to the necessities and solve problems such as:

- Management of complexity.
- Fierce competition between leading sea ports due to the high number of companies using maritime transport routes.
- Ability to handle the maximum volume of containerized cargos.
- Highly fragmented industry with a large number of agents.
- Compliance with international security and safety requirements.
- Need for accurate information for planning and operations.

It was necessary to meet the requirements of agents and users. Users need simplified business processes which do not force them to make changes to the systems they use and do not change the nature of their relationships with their partners.

12.2.2 The Solution

Valenciaportpcs.net is the result of a wide-ranging process aimed at improving the productivity of the port community using IT. The IT web platform developed to cater to the needs of the port of Valencia is a further development and improvement of the SIC (Community Information System) and EDI-based applications for the exchange of information between the agents of the port community and the Port Authority of Valencia.

Valenciaportpcs.net aims to establish itself as the main tool supporting the operations between said agents by:

- Modernizing logistics management: employing standardized information exchanges to facilitate business and administrative procedures and operations relating to cargo movements at Valencia ports.
- Optimizing seaport-land integration: by developing the existing integration between the port community and land transport services on both the "land" side (with haulage operators) and on the "sea" side through integration with the world's leading carriers via INTTRA and GT Nexus. This means that users are able to view the entire transport chain on the website.

- Interconnecting associated ports: making integration options with other associated ports more accessible, thereby facilitating operations and the exchange of information between port communities.

The fact that valenciaportpcs.net covers business and transactions for sea, port and land operations constitute yet another step forward in the evolution of port information systems.

The new IT system integrates all the agents involved in the logistics operations that take place in the port and solves the coordination problems they have in the area of loading and unloading of shipments. The tool provides information relating to logistics processes like:

- Sea: sending and confirmation of documentary procedures relating to transactions, confirmation of booking and shipping instructions.
- Port: the electronic processes involved in port of call requests, dangerous goods declarations, pre-loading cargo manifests and summary declarations.
- Land: processing of haulage orders, cargo acceptance and delivery orders as well as rail loading orders.

In addition to administrative services, valenciaportpcs.net also offers features of a more operational nature. These operational and logistics services include:

- Integral track and trace: this service enables platform users to obtain track and trace information about their shipments such as the current status of their cargo, transshipments carried out or documents processed. It also allows them to integrate this information in their systems to make it available to their customers. In addition to being accessible from the client application and via electronic messages, this service is available in the private section of the valenciaportpcs.net web page, enabling users to obtain information about their shipments anywhere and at any time.
- Quality control: this service provides information from the different agents involved in transportation about compliance with the guarantees offered by the Port of Valencia Seal of Quality Guarantee and also the quality management provided by users of the platform themselves. A series of activity reports has also been designed to provide graphs of the volume of use of each service and the number of transactions carried out via valenciaportpcs.net

12.2.3 Challenges

Challenge 1: Integration of all the agents operating at the port.

Challenge 2: The necessity to create a new application to integrate the IT systems used by the agents with the new web service in order to transfer and include the information provided by these agents.

Challenge 3: The lengthy deployment and acceptance process does not begin until the software has been implemented, which means it takes a long time for the port community to recognize the value of the new system.

Challenge 4: The port of Valencia is a public body; this means that, in order to implement this project, the PAV had to tackle administrative, legal and procedural issues that resulted in added complexity in the software development process.

12.2.4 *Lessons Learnt and Success Factors*

- Lesson 1: Integration of all the agents operating at the port improves productivity and helps to boost efficiency.
- Lesson 2: Error-free processes help to reduce costs.
- Lesson 3: Integrating different systems and making sure they run smoothly can ensure that information is provided faster and more reliably.

12.2.5 *Benefits*

Easy access to all integrated logistics information (sea, port and land), improved performance of operations (which improves quality and customer service), efficient transactions, one-stop access and communication with carriers are the key features of the modern design of port of Valencia. The system is easier to use thanks to two methods of integration: electronic messaging integrated in user applications and an exclusive customer application.

12.2.5.1 Economic

- The port is now more efficient and flexible, customer service has improved, and agents can handle more TEUs. Port services have improved and information is provided faster and with greater reliability.
- Productivity has improved due to standardization of all the agents; moreover, truck drivers and the remaining the agents have also increase their productivity because information is more precise and waiting times and errors have been reduced.
- This improves processes and reduces the amount of paperwork. In addition, the availability of more precise data reduces excessive costs. Optimization of processes also has a positive impact on costs.

12.2.5.2 Environmental

- As information is more reliable, waiting times and the number of the trucks have been reduced or optimized, and exhaust emissions are also lower.

- Whenever a port increases the scope of its operations and the number of containers moved, it is generally necessary to build more docks and buildings. This need has been reduced thanks to optimized utilization of the existing dock facilities.

12.2.5.3 Social

- Reducing waiting time at terminals and less congestion of ships at the port etc. decreases hazardous emissions.
- Noise is reduced due to the lower number of trucks at the port and reduced congestion.

12.2.6 Transferability

12.2.6.1 Transferables

This technique can be used in any industry because consists of integrating and communicating all the agents involved in the supply chain. They only need a common way of communication (for example EDI).

In general this way of working can be used in any kind of geography but is very relevant in ports because of the distance between the different agents. There is no problem in using this technique for little companies but sometimes it can be too expensive.

12.2.6.2 Limitations

There are no limitations in terms of efficiency and integration of the companies. The problems begin when a small company wants to use this kind of technology. As mentioned before, companies need to invest a lot of money in order to achieve this level of technology and they are sometimes unable to do so.

12.3 Intermodal Terminal: Lovosice – Inter-Modal Terminal

Miroslav Rumler

The INTER-MODAL terminal in Lovosice, CZ, was launched as a project of Ceske drahy a.s. and DUSS mbH. The terminal offers public access to services such as swap body handling, container handling and road trailer handling between road and rail as well as pallet and parcel handling inside a modern warehouse. The terminal broadens the portfolio of services offered by Ceske drahy a.s. and it also fills the gap in the intermodal transport of road trailers from the Czech Republic to the German inland ports.

12.3.1 Problem Description

In the Central European region, road and rail transport have traditionally been competing for the vast majority of hauls exceeding distances of around 300 km. FTL shipments to and from major European sea ports in particular have been the subject of fierce competition between these two modes of transport.

CD Cargo a.s., a subsidiary of Ceske drahy a.s., is one of the top six railroad companies in Europe. It offers wide a range of services in the intermodal transport of goods. In 1994, together with the private Kombiverkehr company, Ceske drahy a.s. launched the Ro-La inter-modal project based on accompanied transport of road trucks. At that time, a special terminal designed for loading and unloading of trucks onto railroad cars was built at the site of the CSKD Intrans container terminal and the Lovosice railroad station.

After EU accession of the Czech Republic, the system become obsolete and the service was finally disconnected. Ceske drahy a.s. was suddenly faced with the problem of how to use the facility. At the time, Ceske drahy a.s. initiated another project with the ultimate aim of generating business for the terminal as well as reducing the volume of road traffic en route from the Czech Republic to the inland German ports. As it is now, the terminal provides public access to all intermodal as well as conventional road-rail transport services, and there are plans for further construction to meet the ever-growing demand from existing and new customers.

12.3.2 The Solution

The intermodal terminal in Lovosice, CZ, is the end result of a project aimed at responding to the inevitable decline of rail transport in favor of road transport in the early nineties. The former balance of road and rail transport in former Czechoslovakia partly ensured by vast shipments of bulk materials across the former Eastern Bloc and partly by government incentives no longer exists. Most commercial entities would prefer road transport for its flexibility, speed, and customer orientation. Newly established industrial sites and logistic parks have been built with no direct access to rail transport services, and no major sums have been invested in existing container terminals either. The Czech rail transport operator – CD – owned by the Czech government has suffered a substantial decline in transported volumes and, consequently, also total revenues.

The solution builds on two fundamental prerequisites: the first one is the fact that the road network along the main cargo streams (incl. the one between Germany and the Czech Republic) are operating close to their capacity limits. The second is the assumption that demands for intermodal transport can be boosted only by construction of a new or major reconstruction of an existing intermodal facility permitting fast, safe and affordable handling of individual cargo units.

CD therefore initiated the establishment of a new service organization called CD Duss, a joint venture of CD and Duss mbH. The newly established organization maintains and develops the intermodal terminal in Lovosice, CZ and, even more importantly, it plays a major role in attracting traditional road freight customers to intermodal solutions by providing superior service, customized solutions and reasonable pricing driven by sufficiently high volumes.

The final solution is in fact a blend of three elementary components: customer orientation, service diversification and the desire for better utilization of existing assets. These three drivers have predefined the nature of the entire project.

Pronounced customer orientation was the most significant driver for the subsequent transformation of the company – or rather its cargo division – into CD in 2007.

Cargo a.s. – development from the traditional concept of a governmental institution into a modern operator with a strong focus on the quality of all services offered. This has also involved ISO 9001 certification by Lloyd's Register Quality Assurance. Customer focus was a key driver of systematic diversification efforts into the provision of logistics services which ultimately resulted in major investment in the Lovosice container terminal. The former terminal was then converted into a modern intermodal terminal offering a wide range of logistics services. The expansion and development of the Lovosice is not yet complete.

Recent observations confirm the original estimates: the solution generates additional business for the parent company, creates new jobs in the present location and, most importantly, has excellent potential to handle a high proportion of the of traditional road cargo on its journey from the Czech Republic (as well as a certain volume of transit shipments) to Germany and vice-versa.

12.3.3 Challenges

Challenge 1: Definition of a suitable portfolio of services meeting the current and future requirements of customers and potential customers

Challenge 2: Departure from the concept of a pure rail transport operator and involvement in the field of customized logistics solutions

Challenge 3: Implementation of new elements into the existing marketing strategy of a traditional provider of rail transport services

12.3.4 Lessons Learnt and Success Factors

- One of the key important challenges of this project was to bring together rail and road transport operators. The main lesson learned was that once benefits for all interested parties were identified, they were willing to participate and make valid contributions to the project.
- Increasingly, environmental concerns are considered to be valid reasons for changing "traditional" arrangements; Especially with line hauls becoming longer in a common European market that is still growing, shippers are more willing to consider "going intermodal".
- Proper intermodal solutions enable operators to make better use of their assets, as well-designed solutions have the ability to generate a critical mass of shipments.

12.3.5 Benefits

The general benefits of Inter-Modal are:

- An increase of the market share of Ceske drahy a.s. in the intermodal business
- Reduced negative environmental impact of road transport along the E 50 between the Czech Republic and German inland ports and
- An increase in the volume of unaccompanied intermodal goods transportation in the continental hauls from and to the Czech Republic

12.3.5.1 Economic

- The solution introduced a new service – unaccompanied intermodal transport
- The solution benefits from the fact that the services of the transport provider and the logistics provider are ultimately offered by the same company. This is the ideal prerequisite for customized solutions for individual customers
- The practice significantly enhanced the availability and capacity of logistics centers; the independent provider guarantees instant flexibility and access for virtually any customer

12.3.5.2 Environmental

- Decreased noise pollution from trucks on the roads
- Helps to decouple the overall increase of transport volumes and reduces traffic congestion on the roads
- Optional third mode available nearby (rail access to the Elbe river port)

12.3.5.3 Social

- Staff have been trained for more qualified jobs, e.g. new warehousing and handling technologies.
- Positive impact on employment in companies connected with terminal operations.

12.3.6 Transferability

12.3.6.1 Transferables

Transferable to any country with an adequate rail and road infrastructure. Limited transferability in terms of sectors. There is no problem in using this technique for small, but it may sometimes be too expensive.

12.3.6.2 Limitations

Except for inadequate infrastructure (incl. modern intermodal facilities) the only limitation we have been able to identify was the sincere desire of two modes of transport, i.e. road and rail, to share the business; the reward for cooperation should be an overall increase in business volumes for both (or all) partners; However, in times when most road haulers are faced with a critical lack of drivers, this limitation should shortly become obsolete and become an opportunity.

12.4 Cargo Securing: Daimler AG – Adaptive Logistics

Erich de Vries and Balkan Cetinkaya

Daimler shows how it uses its own, "adaptive logistics" strategy based on a legal regulation for load securement to pinpoint and implement potential for process optimisation. The solution shows that requirements imposed by society can lead to efficient and sustainable processes through the use of innovative and cooperation-based measures.

12.4.1 Problem Description

In light of the changed legal framework resulting from the implementation of the German Transport Reform Act (TRG), the introduction of new technology for trailers (e.g. curtainsides) and new packaging materials, Daimler AG saw the need to examine and rethink the issue of load securement for material transports.

Two systematic options for load securement on trucks presented themselves: friction-locking securement (tying down with lashing straps) or mechanically interlocking securement (load retention by the body of the vehicle).

12.4.1.1 New Legal Requirements

The main driver was compliance with the legal requirements relating to load securement measures in Germany (derived from the VDI 2700 standard as an acknowledged "rule of technology"). In this area, the legislator used the German Transport Reform Act (TRG) in 2000 to once again assign greater and more clearly defined responsibility for safe load securement to the loading companies as the clients who place the actual orders with the transport companies.

12.4.1.2 Reduction of Process Times

A further objective was to minimise in-plant process time for loading and unloading operations to ensure that the forecast increase in transport volumes did not lead to higher throughout times.

12.4.2 The Solution

The "Daimler Load Security 9.5" directive enables all process participants (material supplier, transport provider and Daimler AG) to secure loads solely by means of the design of the vehicle body and using standardised load carriers.

12.4.2.1 Specifications for Trailer Design

For this purpose, the trailers must meet certain specifications in terms of stability and design. These specifications are as follows:

- Front wall: Reinforced front wall with a continuous width of at least 2.40 m
- Side wall: Pallet stop bar, pallet posts, load securing tarpaulins, blocking boards
- Roof: Lifting roof with reinforced roof design
- Rear portal: Minimum strength and minimum number of locking gears (twist locks)

If they meet these requirements, the trailers can be universally used by other industrial sectors and for other goods (e.g. paper, palletised goods etc.).

12.4.2.2 Requirements for Load Carrier Units

The new load carriers possess dimensions based on the ISO standard and are stackable in mechanically interlocking mode with a footprint of at least 600 × 1,000 mm.

12.4.2.3 Expert Opinion on Compliance with Statutory Stipulations

Daimler continue to develop practical load securement measures for heavy unit loads as required by the authorities within the framework of the German Transport Reform Act (TRG). DEKRA is advising Daimler AG as an independent and recognised expert organisation for load securement.

12.4.2.4 Integration in the Overall "Adaptive Logistics" Management Concept

In line with its overall logistics strategy, Daimler aims to introduce adaptive logistics processes and standards.

One of the key elements in this strategy is the development of stable processes that remain stable even when framework conditions change.

The Daimler Load Security 9.5 directive complies with the statutory provisions and ensures a robust process for efficient and safe transport operations.

12.4.3 Challenges

- Challenge 1: Within the general context of load securement: identification of all stakeholders: plant logistics, police, transport industry, German Federal Office for Freight Traffic (BAG), work health and safety, freight purchasing, legal issues, suppliers, the Association of German Engineers (VDI), bodies and committees, project organisation and project budget.
- Challenge 2: Development of suitable practical tests.
- Challenge 3: Identification of independent testing bodies with suitable experts.
- Challenge 4: Ensuring technical implementation by the body manufacturers (market availability).
- Challenge 5: Extensive training network and long-term implementation by all process participants.

12.4.4 Lessons Learnt and Success Factors

- Shortcoming: Lack of load securement specifications for Daimler load carriers
  - Solution: Creation of a Group-wide working group with independent expert support, definition of a budget

- Shortcoming: Use of pallets and load carriers that are unsuitable for mechanically friction-locking load securement (tying down)
  - Solution: Mechanically interlocking load securement through stabilisation of the vehicle body
- Shortcoming: Lack of load securement specifications for curtain side and MEGA trailers with no side wall for loads
  - Solution: Road tests and stipulation of body design – e.g. pallet stop bar and roof reinforcement
- Shortcoming: Lack of legal predictability for the organisation and particularly for management personnel (shipper responsibility)
  - Solution: Compilation of a guideline with Group-wide validity

12.4.4.1 Key Success Factors in the Implementation of "Daimler Load Securement 9.5"

- Clear-cut identification of stakeholders and their needs
- Retention of mechanically interlocking load securement concept, hence optimisation of process times
- Standardisation of practices through definition of the directive as a corporate guideline
- Clearly defined requirement profile for transport service providers and trailer body manufacturers
- Commissioning of a neutral expert organisation
- Stipulation of tasks, expertise and responsibility in the transport process for load securement
- Legal predictability for employees and management executives
- Front-to-end information and training for process owners including body manufacturers

12.4.5 Benefits

Daimler achieves several benefits based on this good practice. This more operational and technical solution enables Daimler to increase the stability and quality of supply operations. The benefits are of a economic, environmental and social nature.

12.4.5.1 Economic

- Less complex load securement measures and processes resulted in optimised transport costs.

- The improvement of the process resulted in a reduction in overall process time and costs (for loading, safety, transport, unloading) compared to the use of a friction-locking load securement solution in line with the VDI 2700 standard.
- As a result, it is possible to absolutely minimise the need for separate transport insurance.

12.4.5.2 Environmental

- Process improvement compared to friction-locking load securement in line with the VDI 2700 standard increased the eco-productivity of vehicles and reduced the use of load securement equipment (e.g. anti-slip matting, lashing straps etc.).

12.4.5.3 Social

- Accidents involving trucks pose a risk to the health of people and, in the case of hazardous products, to the environment.
- With "Daimler Load Security 9.5", Daimler AG underlines its social responsibility for driving, loading and unloading personnel as well as other road users. Responsible securement of the load is a social necessity in order to ensure the safety and protection of personnel who are directly or indirectly involved in the process. With its example, Daimler shows how a logistics strategy identifies the efficiency potential of a socially driven legal regulation and succeeds in implementing this regulation in the form of high-quality and stable processes.

12.4.6 Transferability

12.4.6.1 Transferables

The adaptations to the trailer equipment for load securement purposes are generally transferable to other industries that have their shipments transported by trucks on the roads.

12.4.6.2 Limitations

Legal requirements can differ from sector to sector and from country to country. This in turn can limit transferability. Different types of loaded goods can also result in additional requirements with regard to products and processes.

About the Authors

Balkan T. Cetinkaya, Master in Business and Engineering (MSc.)

Balkan Cetinkaya (born 1975) studied Business Engineering at the University of Technology in Berlin and at the Berkeley University, Haas School of Business mainly specialised on Logistics, Supply Chain Management, Organisation, and Leadership until 2001.

Afterwards Balkan was engaged as an international Consultant at KPMG Consulting in international projects in Singapore, USA, and across Europe from 2001 to 2005.

Starting 2006 at the Berlin Institute of Technology Balkan managed the European Commission project "bestLog – Best Practice Logistics" until 2010. He is also the project manager of one of Germany's largest studies on Trends and Strategies in Logistics, run regularly by the German Logistics Association (BVL) since 1988. Since 2008 he is doing his PhD about Sustainable Supply Chain Management.

As Balkan emphasizes the alignment of research with practice, he is further on active as a Consultant and Lecturer at ITCL (International Transfer Center for Logistic) and advices several companies and governments in Germany.

Research interest: Green Logistics and Supply Chain Management, Energy Efficiency in Transport and Logistics, Supply Chain Strategy, Balanced Scorecard and Performance Measurement.

Balkan Cetinkaya
Berlin Institute of Technology (TU-Berlin)
Chair of Logistics - H90
Strasse des 17. Juni 135
10623 Berlin
Germany
Email: cetinkaya@logistik.tu-berlin.de;
btcetinkaya@gmail.com

Dr. Richard Cuthbertson, Senior Research Fellow, Saïd Business School; Research Director, Oxford Institute of Retail Management; University of Oxford; Fellow of Green Templeton College, Oxford.

Before joining Oxford to focus on the retail and distribution sector, Richard was a Senior Lecturer at Bournemouth University, and before that held managerial posts in marketing and logistics in the automotive and energy sectors, at Unipart Group of Companies and British Gas.

Richard teaches courses in retail marketing and distribution, focused on developing people, leadership teams, and organisations to improve their capability in delivering a consumer-facing strategy for today and the future. The winner of the Pegasus Prize for eBusiness Future Insights and a founder-contributor to the key text, Retail Strategy: The View from the Bridge, he has published over 80 articles.

Research interests: focus on how policy makers, retailers, manufacturers, and logistics service providers can work together to develop a sustainable consumer society, where customisation through information can be efficiently leveraged in a mass market to provide a highly relevant customer experience.

Dr. Richard Cuthbertson
University of Oxford
Oxford
OX1 1HP
UK
Email: richard.cuthbertson@sbs.ox.ac.uk

Graham A. Ewer, CB CBE FCILT

Born in 1944, Graham Ewer was commissioned into the British Army in the Royal Corps of Transport in 1965. He served a full career as a military logistician, leaving the British Army in 1999 with the rank of Major General. Amongst other things, he took part in the first Gulf War in 1990–1991 as the senior logistics officer of 1 (UK) Armoured Division, played a major role in the formation of a single logistics Corps for the British Army, The Royal Logistics Corps, and served as the Army's Director of Logistics Planning . Later, as Assistant Chief of Defence Staff (Logistics) in the rank of Major General he played a major role in the UK's Defence Review in 1998, and in the creation of the Defence Logistics Organisation, a new organisation providing logistic support for all three British Armed Service. On leaving the British Army in 1999, he became the Chief Executive of the then newly formed Institute of Logistics and Transport, now the Chartered Institute of Logistics and Transport (UK), leading that body through a period of substantial development and investment. On completion of his term in that role, in 2004 he was elected President of the European Logistics Association, ELA, an international not for profit association of over 30 professional bodies across Europe active in logistics management. Whilst in office he saw through a programme of improvement and greater participation in relevant European affairs before standing down as the Association's Immediate Past President in 2007. He was elected an Honorary Member of ELA in 2008. He remains active as an independent consultant in logistics and defence matters.

Research interest: Supply and Value Chain Strategies and Practices; the role of supply chain best practice; logistics education and skills; military logistic systems; the relationship between public transport and freight transport.

Graham A. Ewer CB CBE
c/o The European Logistics Association
Avenue des Arts/Kunstlaan 19
B1210 Brussels
Belgium
Telephone: +32 2 230 02 11
Facsimile: +32 2 230 81 23
Email: ela@elalog.org

Dr. Thorsten Klaas-Wissing, Dipl.-Kaufmann, Vice-Director

Born 1967, vocational training as industrial clerk, studies in Business Administration with majors in Strategic Management, Logistics, Operations Management and Economic Geography at the University of Cologne. From 1996 to 2002 PhD student, research and teaching assistant at the Department of Management and Logistics of the University of Cologne. Winner of the German Science Award 2003 with doctoral thesis "Logistik-Organisation". From 2002 to 2005 consultant, senior consultant and freelance consultant at Droege & Comp. GmbH, Düsseldorf, Simon-Kucher & Partners, Bonn and Dr. Klaas Logistik-Beratung, Cologne. 2006 project manager at the Kühne-Institute for Logistics, University of St.Gallen, Switzerland. Since 2007 Vice Director Chair of Logistics Management, University of St. Gallen.

Research interest: Supply Chain Management, Supply Chain Strategy, Logistics-Organization, Logistics Service Provider, Contract Logistics Logistics Networks.

Dr. Thorsten Klaas-Wissing
University of St.Gallen
Chair of Logistics Management LOG-HSG
Dufourstrasse 40a, CH-9000 St.Gallen, Switzerland
Tel. +41 71 224 7299
Fax: +41 71 224
Email: thorsten.klaas@unisg.ch, http://www.logistik.unisg.ch

Dr. Wojciech Piotrowicz, Research Fellow at the Saïd Business School, University of Oxford.

He has experience as a researcher and project manager of EU-funded and international research projects gained at the Maritime Institute in Gdańsk, as well as at University of Gdańsk (Poland), Kent (Canterbury, UK) and Oxford. Wojciech teaches Operations and Technology management at Saïd Business School. Before joining academia he was managing department at ICT company and worked as IT consultant.

His research specialisation and expertise are in the areas of performance measurement and evaluation, especially in the context of supply chains and information systems.

Dr. Wojciech Piotrowicz
University of Oxford, Saïd Business School
Oxford OX1 1HP
United Kingdom
Email: wojciech.piotrowicz@sbs.ox.ac.uk

Christoph Tyssen, Dipl.-Wirtschaftsingenieur, Reseach Associate

Born in 1979, studies of Industrial Engineering at the University of Karlsruhe (TH), Germany and the Universitat Politècnica de Catalunya in Barcelona, Spain. His majors were Strategic Management and Organizations, Logistics and Supply Chain Management as well as Marketing and Entrepreneurship. Since 2006 he works as a Research Associate at the Chair of Logistics Management, University of St. Gallen.

Research interest: Supply Chain Management, Third Party Logistics Providers, Contract Logistics, Knowledge and Innovation management

Christoph Tyssen
University of St. Gallen
Chair of Logistics Management LOG-HSG
Dufourstrasse 40a, CH-9000 St. Gallen, Switzerland
Tel. +41 71 224 7285
Fax: +41 71 224 7315
Email: christoph.tyssen@unisg.ch, http://www.logistik.unisg.ch

Project Partners by Institution and Name

Institution	Name
Berlin Institute of Technology	Frank Straube
Berlin Institute of Technology	Peter Franke
Berlin Institute of Technology	Balkan Cetinkaya
Berlin Institute of Technology	Regina König
Chalmers University of Technology	Gunnar Stefansson
Chalmers University of Technology	Kenth Lumsden
European Logistics Association	Nicole Geerkens
European Logistics Association	Graham Ewer
University of St. Gallen	Thorsten Klaas-Wissing
University of St. Gallen	Christoph Tyssen
METTLE	Marielle Labrosse
METTLE	Carmine Biancardi
METTLE	Jean Francois Reynaud
Reliant	Miroslav Rumler
Reliant	Mirek Rumler
Saïd Business School University of Oxford	Richard Cuthbertson
Saïd Business School University of Oxford	Gerd Islei
Saïd Business School University of Oxford	Wojciech Piotrowicz
Itene Packing Transport & Logistics Research Centre	Emilio Gonzáález
Itene Packing Transport & Logistics Research Centre	Sonia Guerola
Itene Packing Transport & Logistics Research Centre	Patricia Bellver
Warsaw School of Economics	Krzysztof Rutkowski
Warsaw School of Economics	Barbara Ocicka
Warsaw School of Economics	Katarzyna Gapska

Members of the Advisory and Communications Board